MASTERING THE
MATH MILESTONE

By
Stacy Otillio & Frank Otillio

- ✓ **Step-by-Step Simple Explanations**
  to guide your child through the process of solving math problems

- ✓ **Writing and Tracing**
  for practice writing and spelling real math terms and numbers

- ✓ **Lots of Practice Worksheets**
  to help solidify your child's understanding of the lessons

EMPOWERING CHILDREN
FOR A SUCCESSFUL FUTURE

Copyright © 2016 - Stacy Otillio & Frank Otillio

All rights reserved.

Table of Contents

Lesson 1
Less Than, Greater Than & Equal 1

Lesson 2
Counting (0 Through 20) 13

Lesson 3
Comparing Numbers 25

Lesson 4
Comparing Amounts 33

Lesson 5
Angles 43

Lesson 6
Lines 51

Lesson 7
Shapes & Dimensions 61

Lesson 8
Regular Polygons 81

Lesson 9
3-Dimensional Shapes 91

Lesson 10
Addition 105

Lesson 11
The Commutative Property 115

Lesson 12
Subtraction 123

Lesson 13
Tables: Rows & Columns 133

Lesson 14
Counting (Up to 100) 143

Lesson 15
Counting by 2's 157

Lesson 16
Counting by 10's 165

Lesson 17
Counting by 5's 173

Lesson 18
Tallying 181

Solutions 191

LESSON 1

Less Than, Greater Than & Equal

LESS THAN, GREATER THAN & EQUAL

Sometimes we want to compare two things and show which one is bigger and which one is smaller. We use a symbol to do this and it looks like a sideways V.

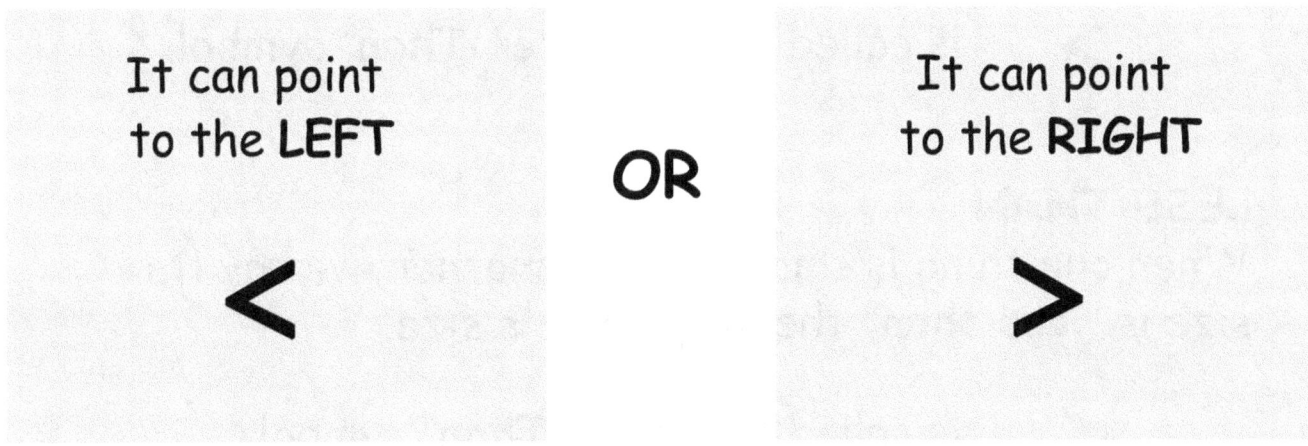

The symbol is drawn with the **big side** facing the **bigger** thing and the **small**, pointy side facing the **smaller** thing.

Trace the symbol below:

The caterpillar on the left is **bigger** than the caterpillar on the right.

Each pointy symbol has a different name, depending on which way it is facing.

GREATER THAN
When one thing is **bigger than** another, we say its size is "**greater than**" the smaller one's size.

> **>** is called the "**Greater Than**" symbol.

LESS THAN
When one thing is **smaller than** another, we say its size is "**less than**" the larger one's size.

> **<** is called the "**Less Than**" symbol.

Example: Compare the size of the stars.
Trace the symbol.

The size of the first star is **less than** the size of the second star.

LESS THAN

Trace the name of the symbol, then write it on your own in the space below:

less than

Trace the "less than" symbols, then write them on your own in the space below:

< < < <

GREATER THAN

Trace the name of the symbol, then write it on your own in the space below:

greater than

Trace the "greater than" symbols, then write them on your own in the space below:

> > > > >

What if 2 things are the same size?

If 2 things are the same size, we say their sizes are **equal**.

= is the "**Equal**" sign.

Example: Compare the sizes of the stars below.
Trace the symbols.

The size of the first star is **less than** the size of the second star:

 <

The size of the first star is **equal to** the size of the second star:

 =

The size of the first star is **greater than** the size of the second star:

 >

EQUAL

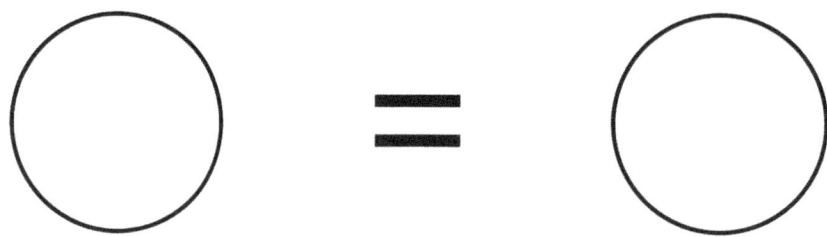

Trace the name of the sign, then write it on your own in the space below:

equal

Trace the equal signs, then write them on your own in the space below:

WORKSHEET
LESS THAN, GREATER THAN & EQUAL

Compare the sizes of the objects below by writing <, > or = in the space between them.

1. 　　

2. 　　

3. 　　

4. 　　

WORKSHEET
LESS THAN, GREATER THAN & EQUAL

Compare the sizes of the objects below by writing
<, > or = in the space between them.

1.

2.

3.

4.

WORKSHEET
LESS THAN, GREATER THAN & EQUAL

Compare the sizes of the objects below by writing <, > or = in the space between them.

1.

2.

3.

4.

WORKSHEET
LESS THAN, GREATER THAN & EQUAL

Compare the sizes of the objects below by writing <, > or = in the space between them.

1.

2.

3.

4.

LESSON 2

Counting (0 Through to 20)

COUNTING (0 THROUGH 20)

We know how to compare sizes, but sometimes we want to compare numbers and amounts. But to do that we will first need to **count**.

Trace the numbers and number names, then write them on your own in the blanks next to them.

0 zero

1 one

2 two

3 three

4 four

Trace the numbers and number names, then write them on your own in the blanks next to them.

5 five

6 six

7 seven

8 eight

9 nine

10 ten

11 through 19

To make the numbers 11 through 19 write a 1 in front of each of the numbers 1 through 9:

11   12   13   14   15   16   17   18   19

.... and next is 20!

Trace the numbers 11 through 20, then write them on your own in the blanks below:

11   12   13   14   15

16   17   18   19   20

Trace the number names in each row, then write them on your own in the blanks next to them.

11 
eleven _____

12 
twelve _____

13 
thirteen _____

14 
fourteen _____

15 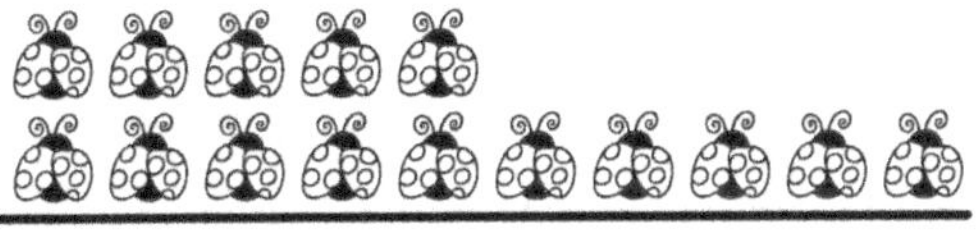
fifteen _____

Trace the number names in each row, then write them on your own in the blanks next to them.

16 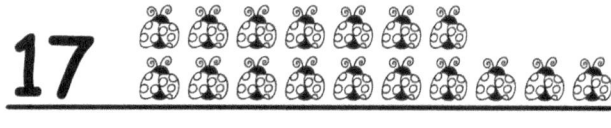 _____

sixteen _____

17 _____

seventeen _____

18 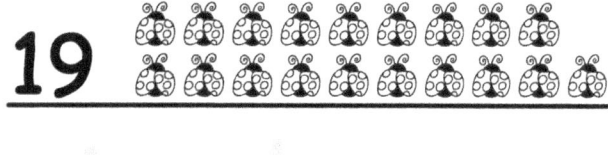 _____

eighteen _____

19 _____

nineteen _____

20 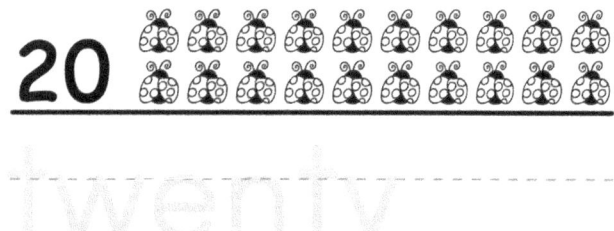 _____

twenty _____

WORKSHEET
COUNTING (0 THROUGH 20)

Count the number of objects in the boxes below, then write the amount in the blanks.

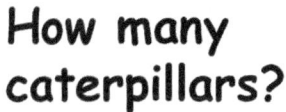

How many caterpillars?

How many rabbits?

How many chicks?

WORKSHEET
COUNTING (0 THROUGH 20)

Count the number of objects in the boxes below, then write the amount in the blanks.

1.
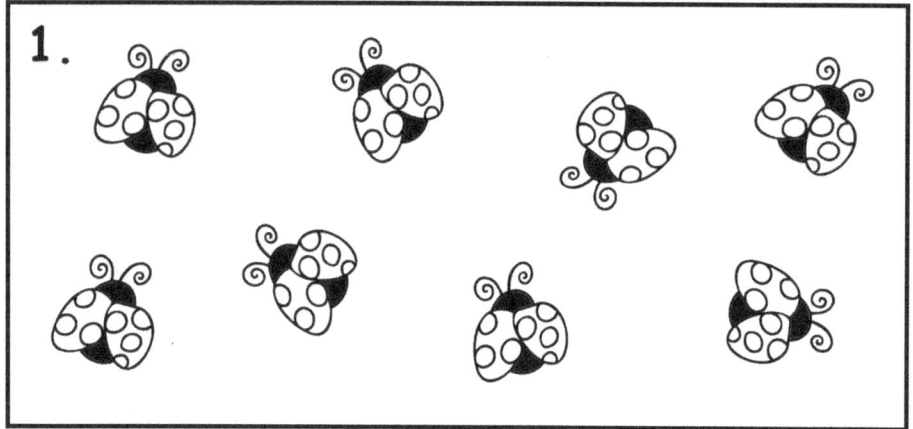

How many ladybugs?

2.

How many monkeys?

3.

How many frogs?

WORKSHEET
COUNTING (0 THROUGH 20)

Count the number of objects in the boxes below, then write the amount in the blanks.

1. How many mice?

2. How many flower pots?

3. How many flowers?

WORKSHEET
COUNTING (0 THROUGH 20)

Count the different objects below, then write the amounts in the blanks for each type.

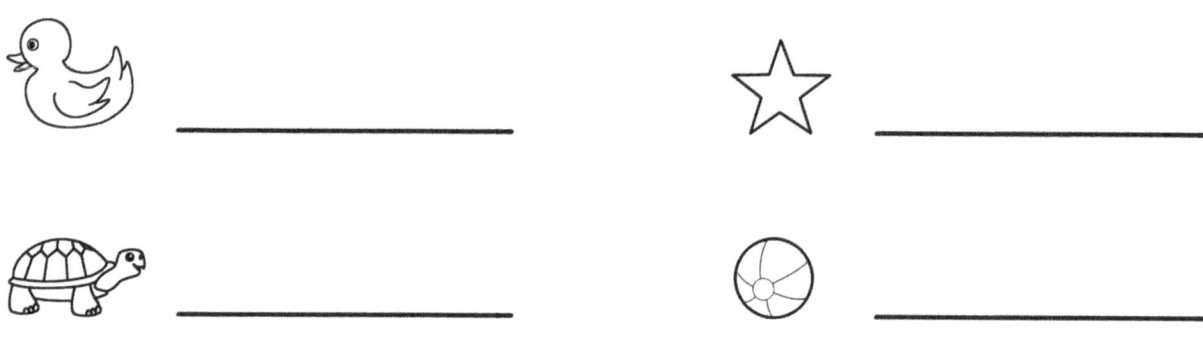

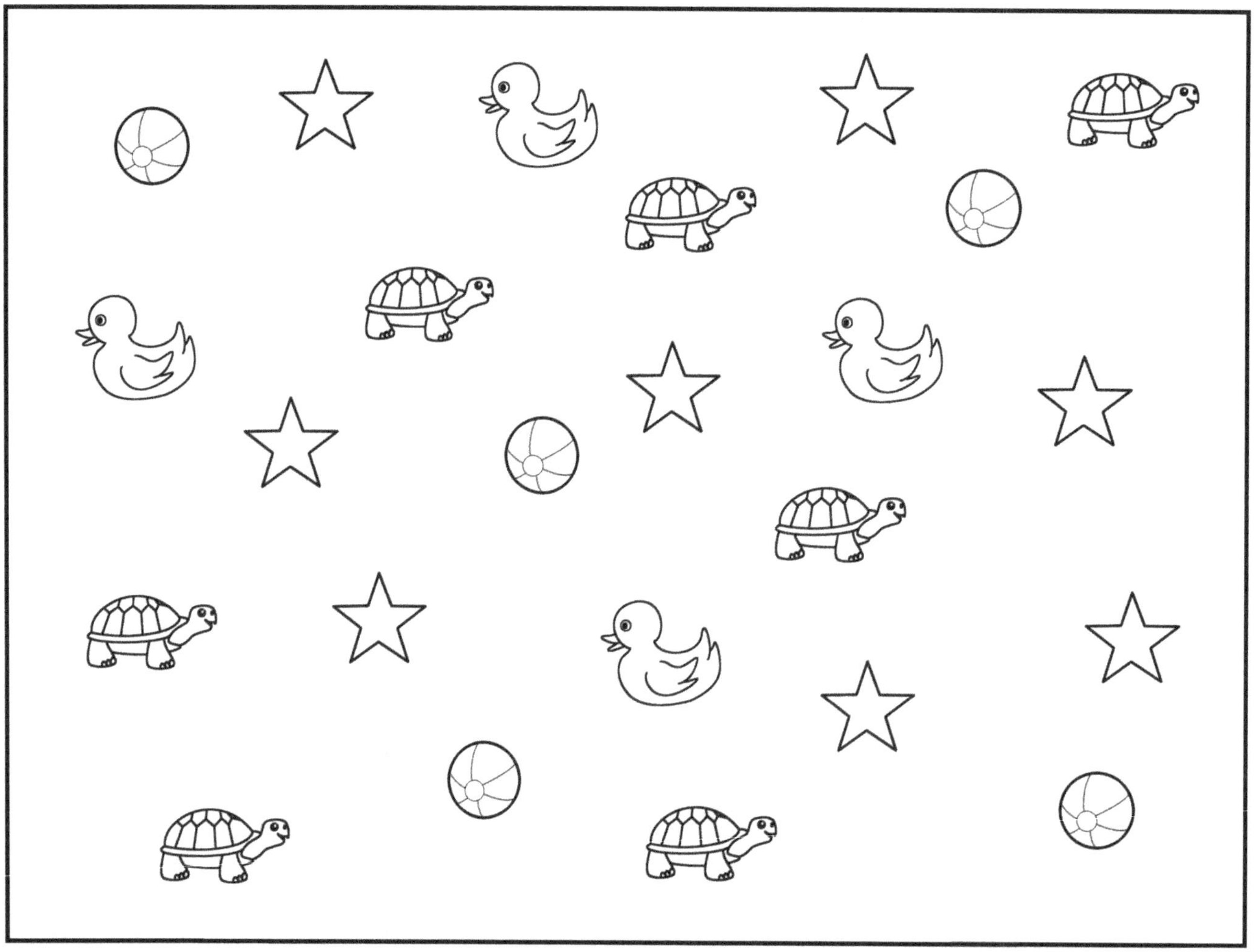

WORKSHEET
COUNTING (0 THROUGH 20)

Count the different objects below, then write the amounts in the blanks for each type.

🪰 _____   🌷 _____

🦉 _____   ✨ _____

LESSON 3

Comparing Numbers

COMPARING NUMBERS

Look at the numbers **1** to **20**:

1 2 3 4 5 6 7 8 9 10 11 12 13 14 15 16 17 18 19 20

As you count **UP** through the numbers **1 to 20**, they get

bigger and bigger and bigger

As you count **DOWN** through the numbers **20 to 1**, they get

smaller and smaller and smaller

Just like you used those pointy symbols < and > to compare sizes, you can also use them to compare numbers.

The little pointy side faces the smaller number and the bigger side faces the bigger number.

Trace the "less than" symbol.

1 < 20

Example: Compare 8 and 10 using < or >.

Find the **8** and the **10** in the numbers **1** to **20** below:

| 1 2 3 4 5 6 7 8 9 10 11 12 13 14 15 16 17 18 19 20 |

8 is before **10**.

This means **8** is **less** than **10**.

Use the symbol with the **small** side facing the **smaller** number (**8**) and the **large** side facing the **larger** number (**10**).

8  10   Trace the "less than" symbol.

Trace the statement below.

8 is less than 10.

Example: Compare 15 and 7 using < or >.

Find the **15** and the **7** in the numbers **1 to 20** below:

1 2 3 4 5 6 ⑦ 8 9 10 11 12 13 14 ⑮ 16 17 18 19 20

15 is after **7**.

This means **15** is greater than **7**.

Use the symbol with the **large** side facing the **larger** number (**15**) and the **small** side facing the **smaller** number (**7**).

15 > **7**   Trace the "greater than" symbol.

Trace the statement below.

15 is greater than 7.

Lesson 3: Comparing Numbers

WORKSHEET
COMPARING NUMBERS

1 2 3 4 5 6 7 8 9 10 11 12 13 14 15 16 17 18 19 20

Trace the numbers and compare them by writing < or > in the boxes between each pair.

3 ☐ 2   12 ☐ 10

5 ☐ 1   5 ☐ 14

6 ☐ 8   11 ☐ 17

2 ☐ 3   10 ☐ 20

7 ☐ 4   12 ☐ 6

WORKSHEET
COMPARING NUMBERS

| 1 2 3 4 5 6 7 8 9 10 11 12 13 14 15 16 17 18 19 20 |

Trace the numbers and compare them by writing < or > in the boxes between each pair.

1 ☐ 3   12 ☐ 14

8 ☐ 5   16 ☐ 8

4 ☐ 7   17 ☐ 20

6 ☐ 3   14 ☐ 15

2 ☐ 7   18 ☐ 12

WORKSHEET
COMPARING NUMBERS

1 2 3 4 5 6 7 8 9 10 11 12 13 14 15 16 17 18 19 20

Trace the numbers and compare them by writing < or > in the boxes between each pair.

2 ☐ 5   11 ☐ 12

4 ☐ 6   18 ☐ 8

1 ☐ 2   15 ☐ 16

5 ☐ 4   20 ☐ 15

8 ☐ 3   17 ☐ 14

LESSON 4

Comparing Amounts

COMPARING AMOUNTS

Now that you know how to compare numbers, let's use that to compare amounts of things.

Example: Let's see which box of cupcakes has the most.

   Step 1: Count the number of cupcakes in each box and write the amounts below each of the boxes.

   Step 2: Compare those 2 numbers and write <, > or = in the blank between the boxes.

1 2 3 4 5 6 7 8 9 10 11 12 13 14 15 16 17 18 19 20

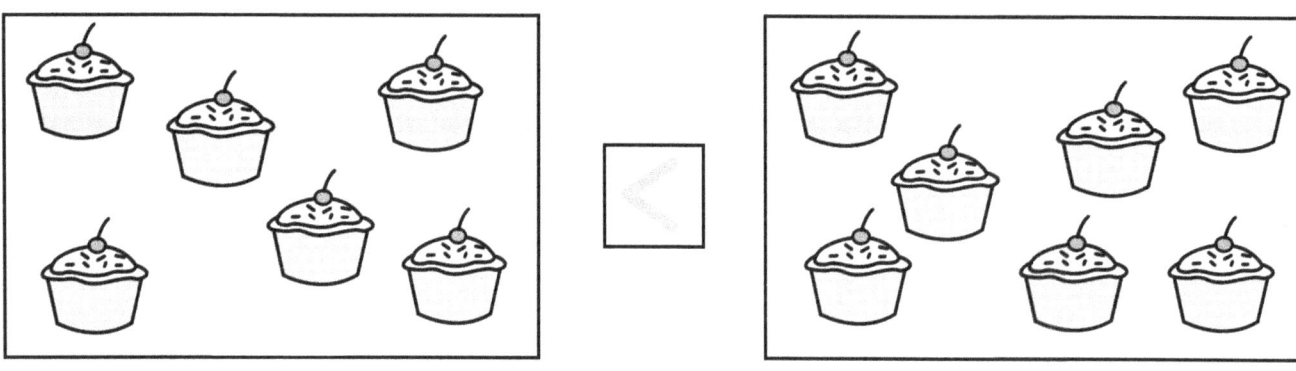

How many cupcakes? 6   How many cupcakes? 7

The amount of cupcakes on the left is **less than** the amount of cupcakes on the right.

Example: Which bowl of goldfish has the most in it?

Step 1: Count the number of goldfish in each bowl and write the amounts below each of the bowls.

Step 2: Compare those 2 numbers and write <, > or = in the blank between the bowls.

1 2 3 4 5 6 7 8 9 10 11 12 13 14 15 16 17 18 19 20

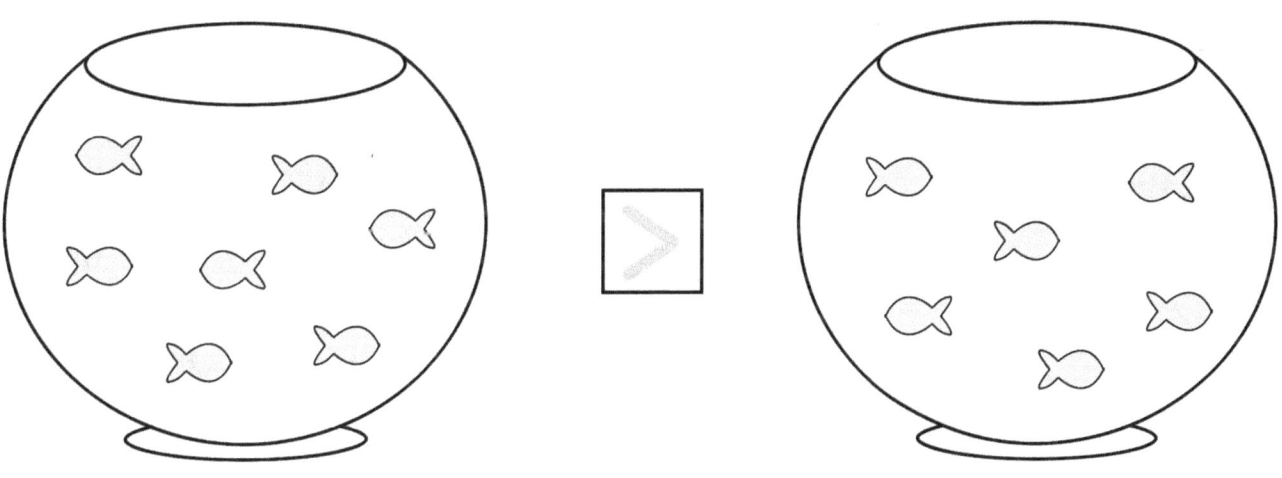

How many goldfish? _7_   How many goldfish? _6_

There are more goldfish in the bowl on the left than there are in the bowl on the right.

WORKSHEET
COMPARING AMOUNTS

1 2 3 4 5 6 7 8 9 10 11 12 13 14 15 16 17 18 19 20

Count the seahorses in each box and write the amount in the blanks below them. Then compare those numbers and write <, > or = in the blank between the boxes.

How many seahorses? _____   How many seahorses? _____

Lesson 4: Comparing Amounts

WORKSHEET
COMPARING AMOUNTS

| 1 2 3 4 5 6 7 8 9 10 11 12 13 14 15 16 17 18 19 20 |

Count the fish in each box and write the amount in the blanks below them. Then compare those numbers and write <, > or = in the blank between the boxes.

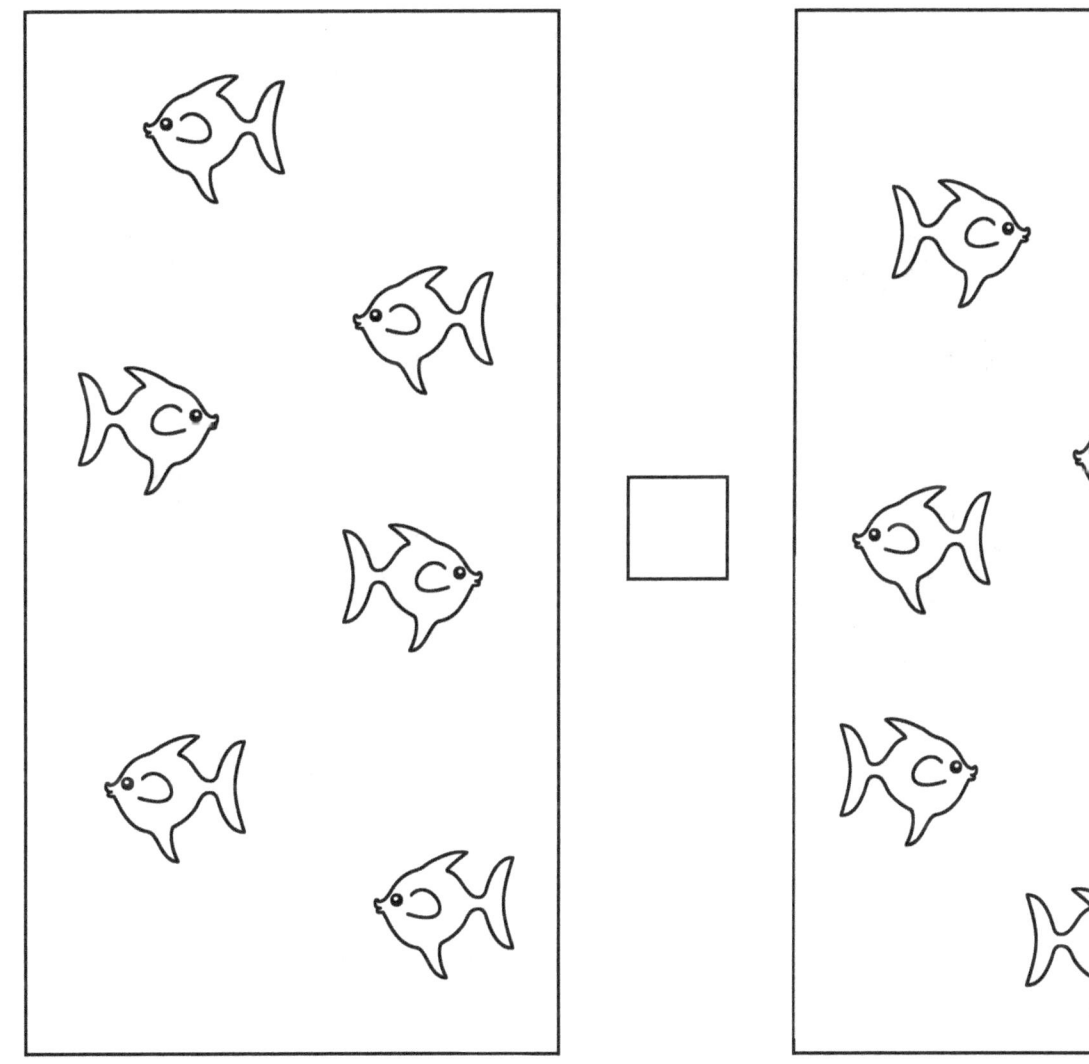

How many fish? _____   How many fish? _____

claymaze.com   Lesson 4: Comparing Amounts   37

WORKSHEET
COMPARING AMOUNTS

1 2 3 4 5 6 7 8 9 10 11 12 13 14 15 16 17 18 19 20

Count the ladybugs in each box and write the amount in the blanks below them. Then compare those numbers and write <, > or = in the blank between the boxes.

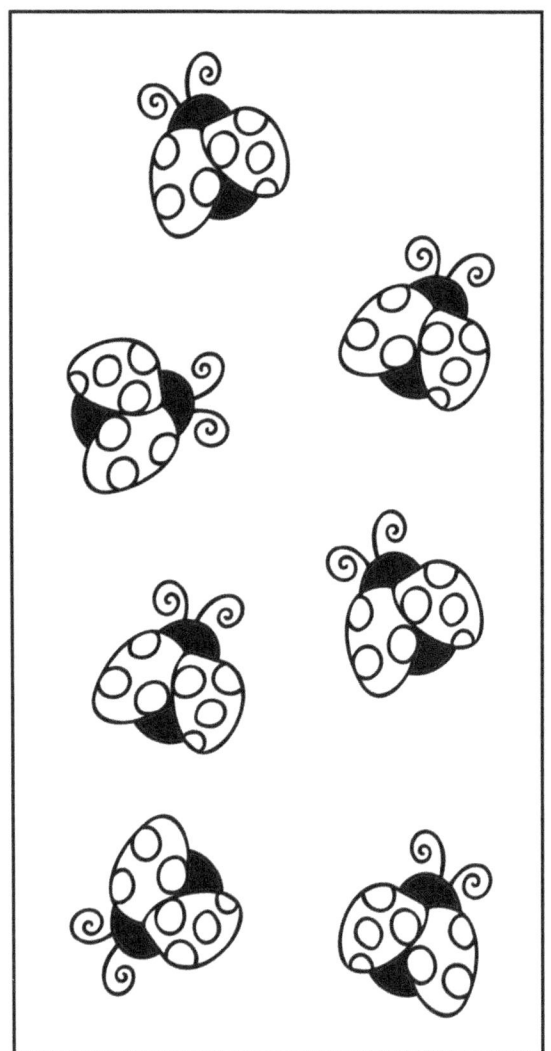  

How many ladybugs? _____   How many ladybugs? _____

38 Lesson 4: Comparing Amounts claymaze.com

WORKSHEET
COMPARING AMOUNTS

1 2 3 4 5 6 7 8 9 10 11 12 13 14 15 16 17 18 19 20

Count the umbrellas in each box and write the amount in the blanks below them. Then compare those numbers and write <, > or = in the blank between the boxes.

How many umbrellas? _____   How many umbrellas? _____

Lesson 4: Comparing Amounts

WORKSHEET
COMPARING AMOUNTS

| 1 2 3 4 5 6 7 8 9 10 11 12 13 14 15 16 17 18 19 20 |

Count the frogs in each box and write the amount in the blanks below them. Then compare those numbers and write <, > or = in the blank between the boxes.

How many frogs? _____   How many frogs? _____

40   Lesson 4: Comparing Amounts   claymaze.com

WORKSHEET
COMPARING AMOUNTS

| 1 2 3 4 5 6 7 8 9 10 11 12 13 14 15 16 17 18 19 20 |

Count the pinwheels in each box and write the amount in the blanks below them. Then compare those numbers and write <, > or = in the blank between the boxes.

How many pinwheels? _____   How many pinwheels? _____

Lesson 4: Comparing Amounts

LESSON 5

Angles

ANGLES

When 2 lines are joined at one end, with the other ends moved away from each other, an ANGLE is formed between them ... like the hands of a clock.

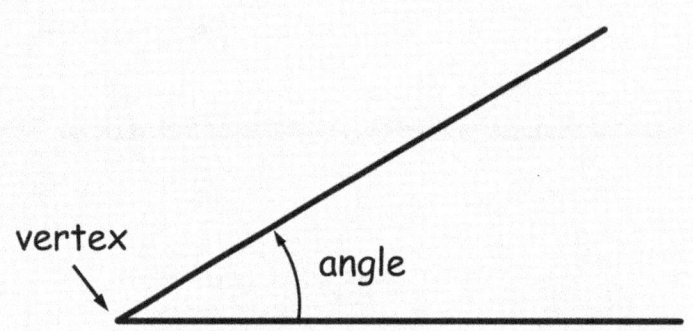

VERTEX
The **vertex** is the point where the 2 lines are connected.

ANGLE
The **angle** is the amount of turn between the 2 lines.

Trace the words, then write them on your own in the spaces below:

RIGHT ANGLE

A right angle looks like the corner of a room or a corner of this page.

RIGHT ANGLE

TRACE THE RIGHT ANGLE

This is a RIGHT ANGLE. It's like the corner of a page or a room.

Trace the angle type, then write it on your own in the space below:

right angle

ACUTE ANGLE

An acute angle is smaller than a right angle.

ACUTE ANGLE

This is an ACUTE ANGLE. It is smaller than a right angle.

TRACE THE ACUTE ANGLE

Trace the angle type, then write it on your own in the space below:

acute angle

OBTUSE ANGLE

An obtuse angle is bigger than a right angle.

OBTUSE ANGLE	TRACE THE OBTUSE ANGLE
This is an OBTUSE ANGLE. It is bigger than a right angle.	

Trace the angle type, then write it on your own in the space below:

obtuse angle

WORKSHEET
ANGLES

Which type of angle is formed by the pairs of lines below? Circle the correct answer.

1.

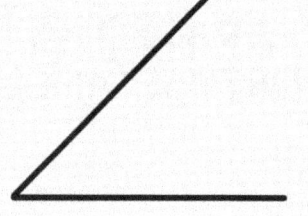

Right   Acute   Obtuse

2.

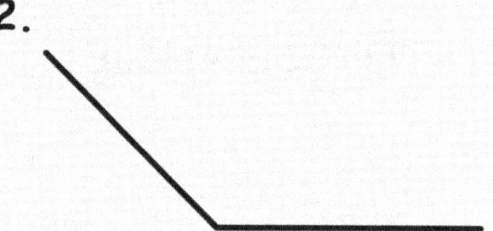

Right   Acute   Obtuse

3.

Right   Acute   Obtuse

4.

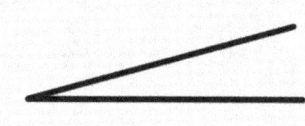

Right   Acute   Obtuse

5.

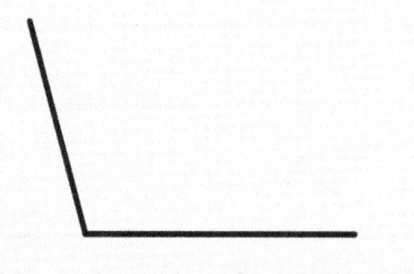

Right   Acute   Obtuse

6.

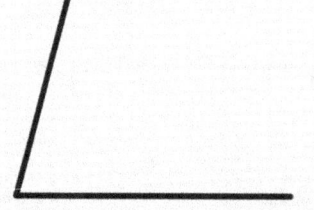

Right   Acute   Obtuse

Lesson 5: Angles

WORKSHEET
ANGLES

Which type of angle is formed by the pairs of lines below? Circle the correct answer.

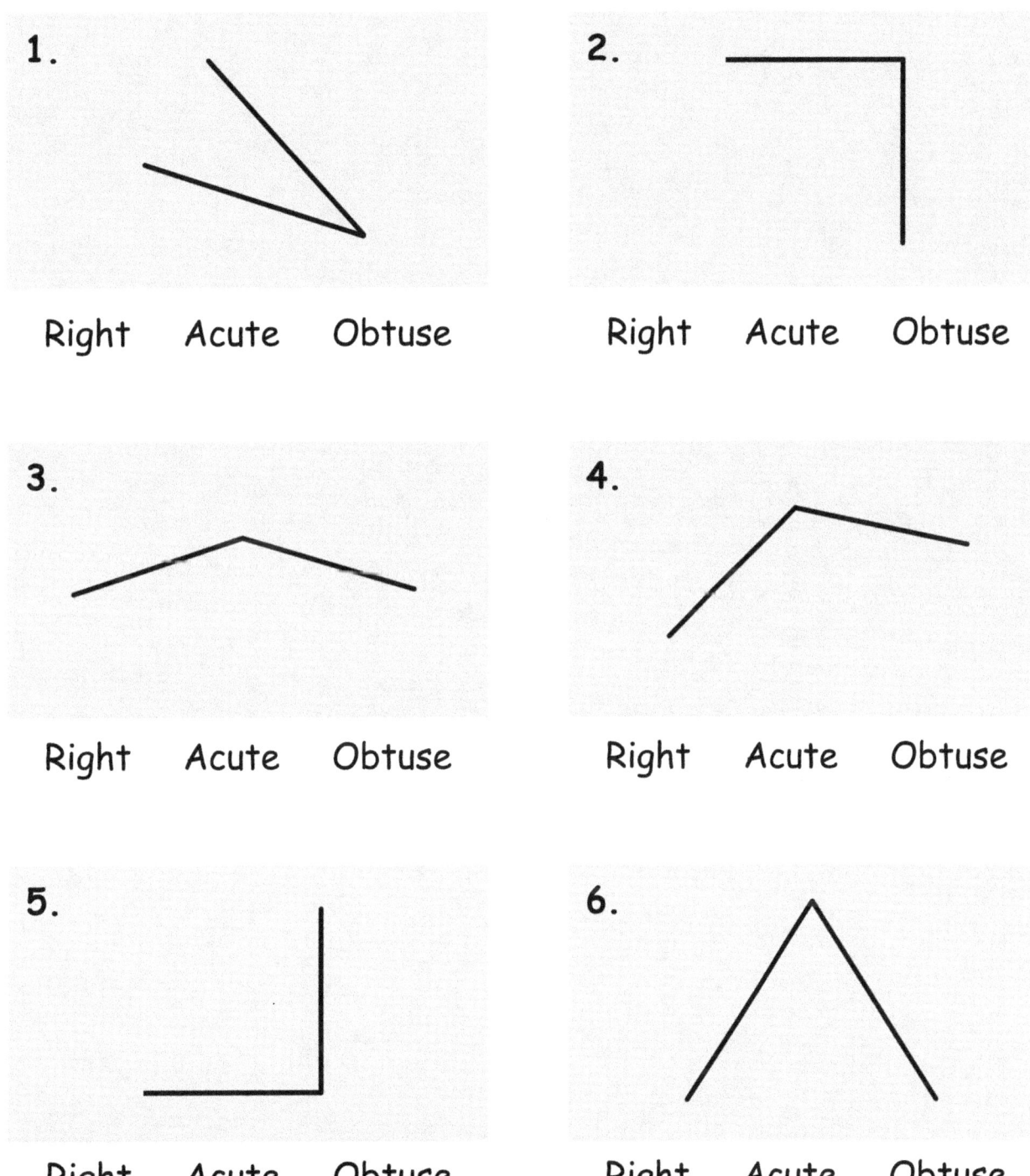

1.

Right   Acute   Obtuse

2.

Right   Acute   Obtuse

3.

Right   Acute   Obtuse

4.

Right   Acute   Obtuse

5.

Right   Acute   Obtuse

6.

Right   Acute   Obtuse

LESSON 6

Lines

LINES

HORIZONTAL

HORIZONTAL LINE
A horizontal line goes from side to side.

VERTICAL

VERTICAL LINE
A vertical line goes up and down.

None of these lines are horizontal or vertical.

Trace the HORIZONTAL and VERTICAL lines below:

HORIZONTAL LINES

VERTICAL LINES

horizontal lines

Trace the words, then write them on your own in the space below:

horizontal lines

vertical lines

Trace the words, then write them on your own in the space below:

vertical lines

INTERSECTING LINES

Lines **intersect** if they cross each other.

These lines intersect:

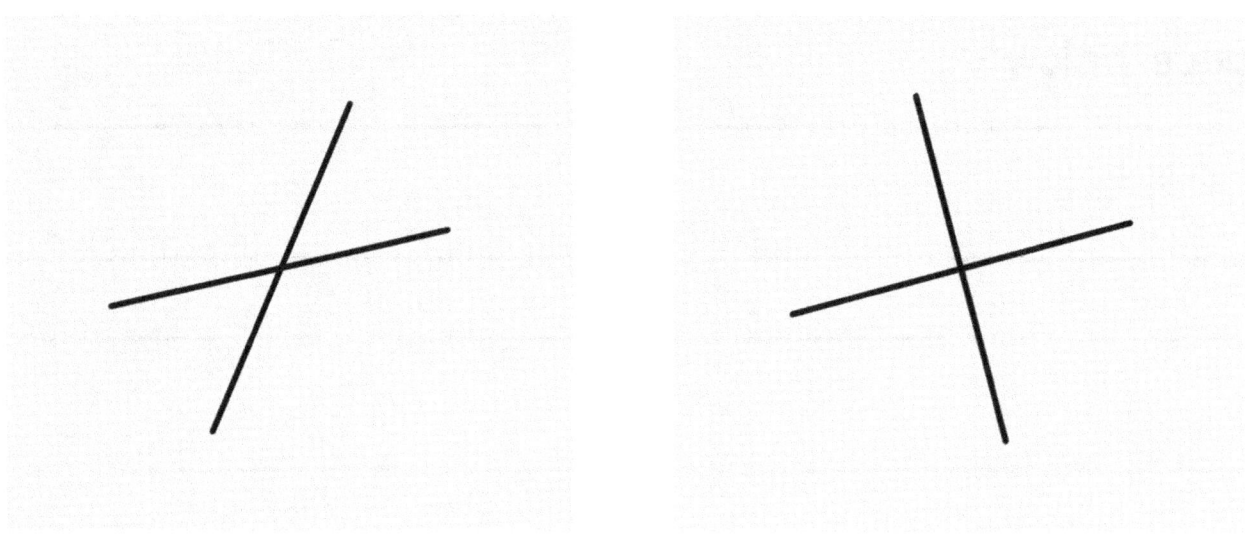

Trace the intersecting lines below:

Draw an intersecting line for each of the lines below:

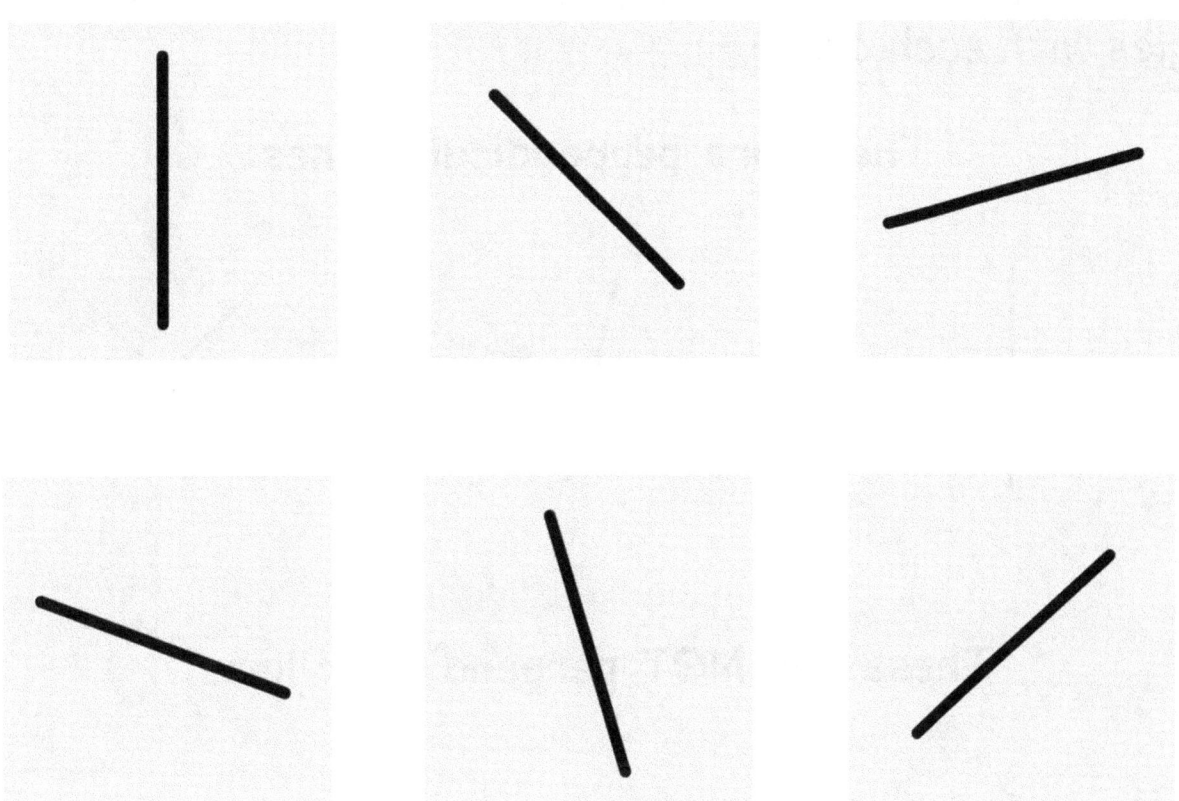

Trace the words, then write them on your own in the space below:

intersecting lines

PERPENDICULAR LINES

Perpendicular Lines are intersecting lines that form right angles with each other.

These are perpendicular lines:

These are NOT perpendicular lines:

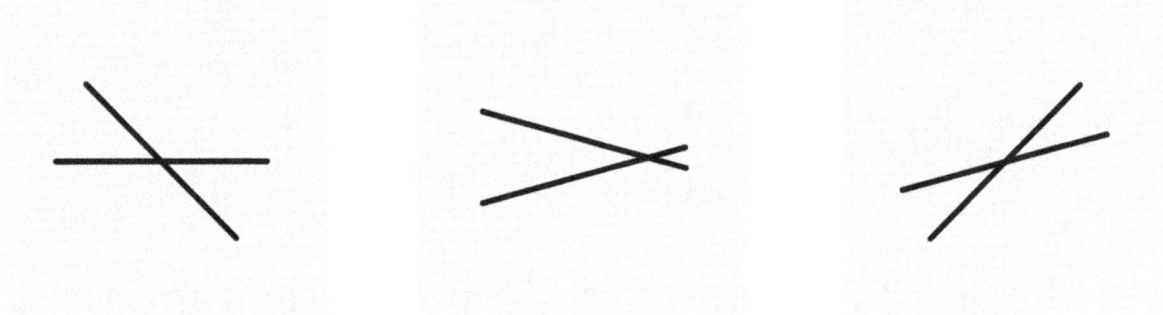

Trace the perpendicular lines below:

PARALLEL LINES

Parallel Lines can never meet no matter how long they are and they are always the same distance apart.

These are parallel lines:

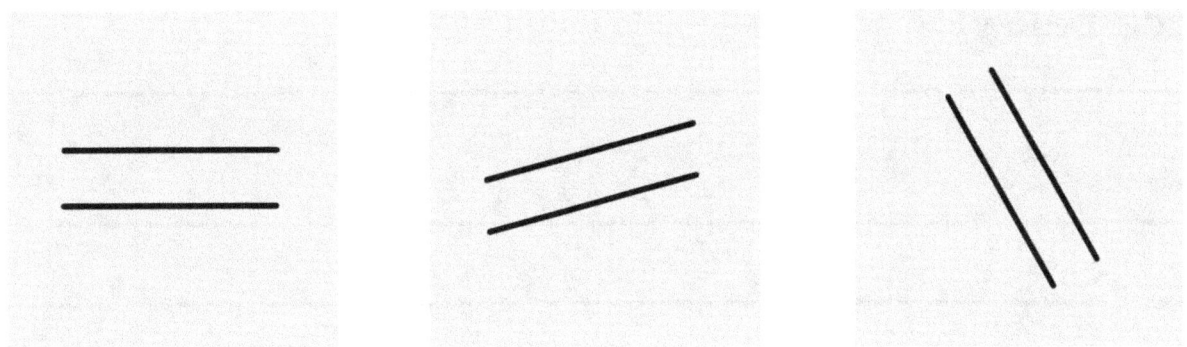

These are NOT parallel lines:

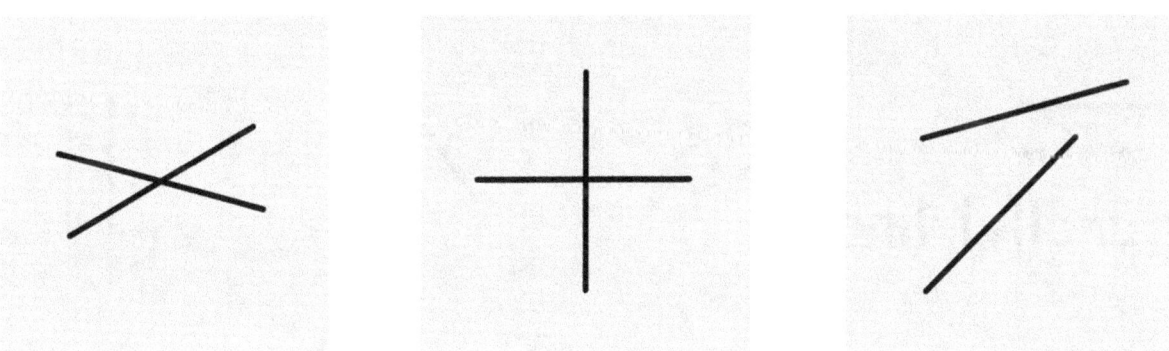

Trace the parallel lines below:

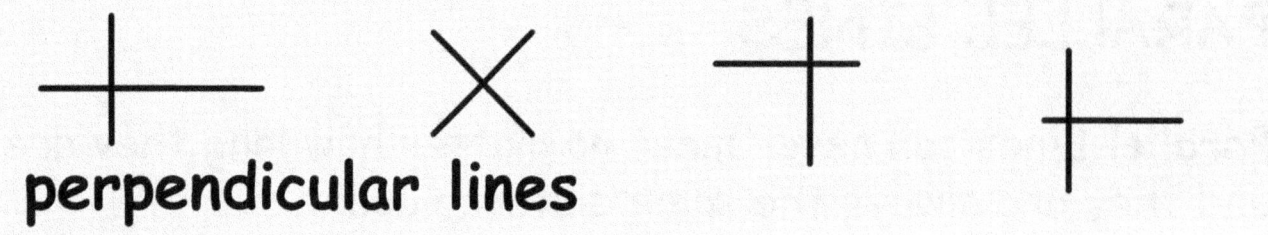

perpendicular lines

Trace the words, then write them on your own in the space below:

perpendicular lines

parallel lines

Trace the words, then write them on your own in the space below:

parallel lines

WORKSHEET
LINES

1. Draw 3 horizontal lines in the space below.

2. Draw 3 vertical lines in the space below.

3. Draw a set of intersecting lines in each of the spaces below.

WORKSHEET
LINES

1. Draw a PERPENDICULAR line for each of the 3 lines below.

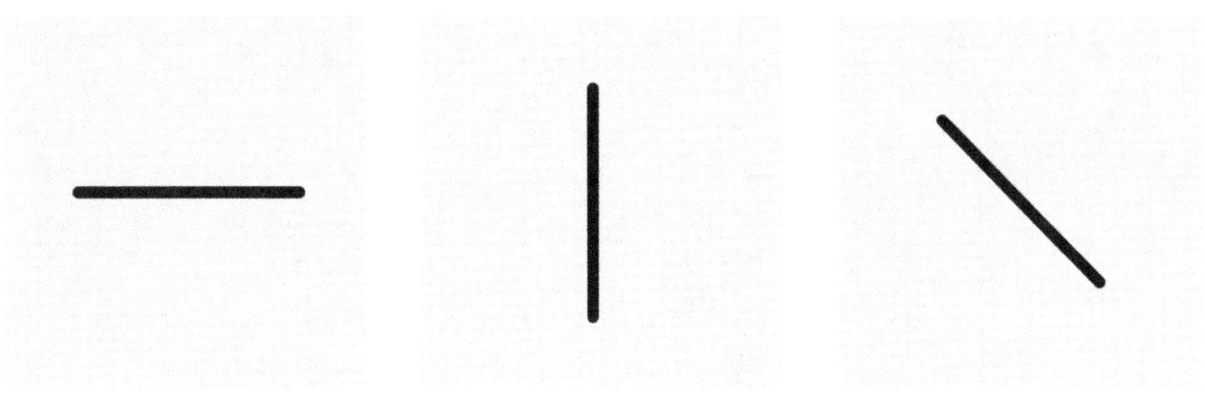

2. Draw a PARALLEL line for each of the 3 lines below.

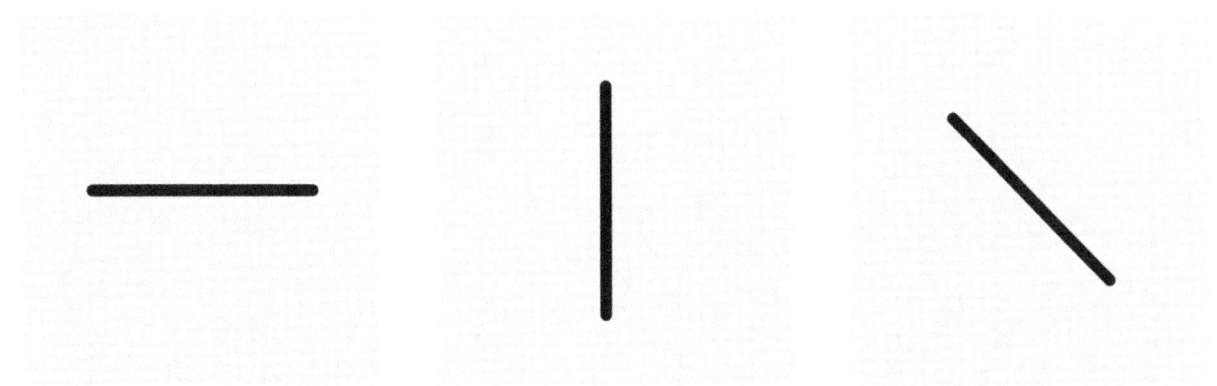

LESSON 7

Shapes & Dimensions

SHAPES & DIMENSIONS

Dimensions are things we measure like length, width and height.

LENGTH

Lines have **1 dimension: Length**
A line's length is a measure of how long the line is.

LENGTH

Trace the words, then write them on your own in the spaces below:

dimension   length

WIDTH AND HEIGHT

2-dimensional shapes are flat shapes and they have 2 dimensions: **width** and **height**.

A shape's **width** is a measure of how **wide** the shape is.
A shape's **height** is a measure of how **tall** the shape is.

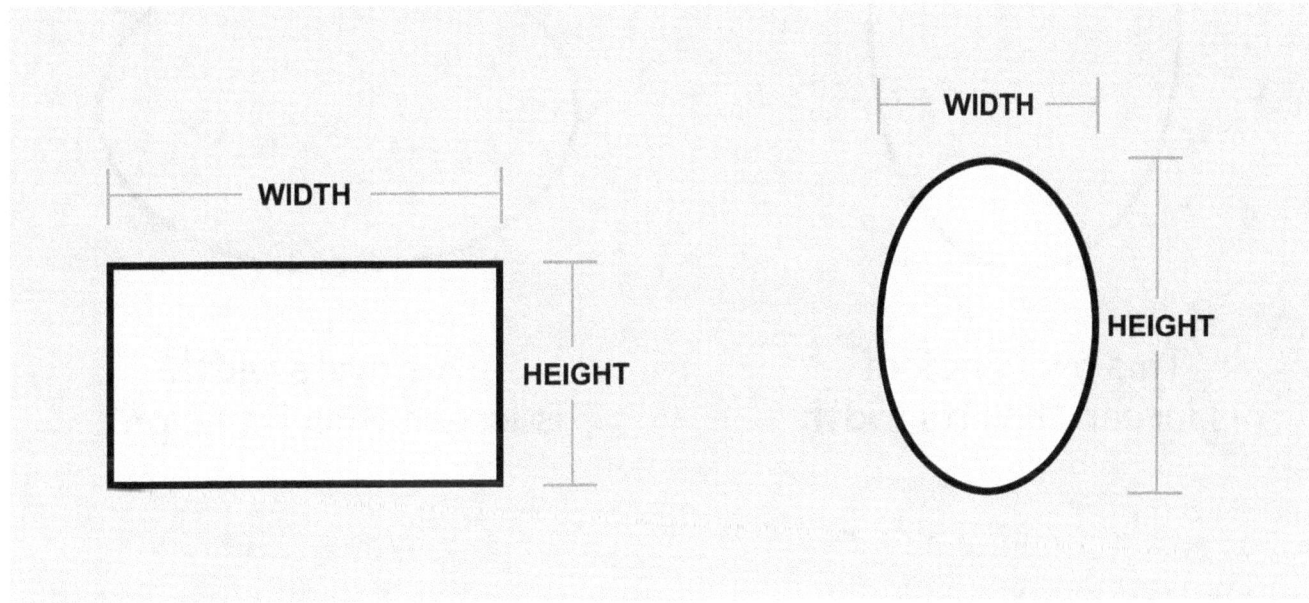

Trace the words, then write them on your own in the spaces below:

width   height

OVALS

Ovals are **oblong** ... like a stretched out circle.

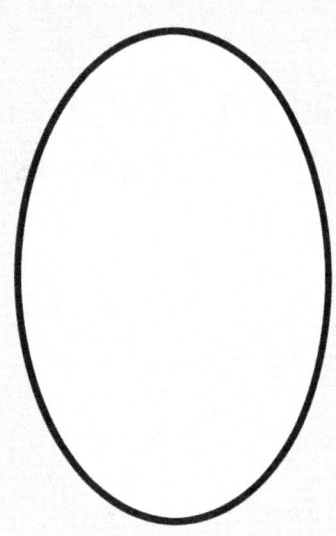
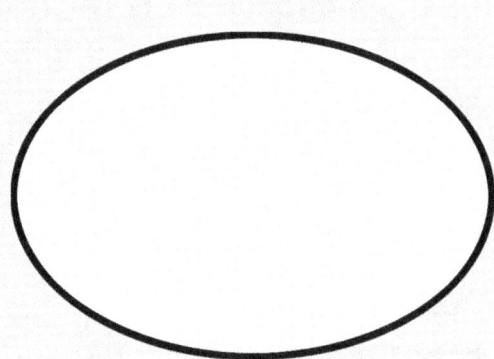

This oval's **height** is larger than its **width**.

This oval's **width** is larger than its **height**.

Trace the ovals below.

CIRCLES

Circles are perfectly **round**.

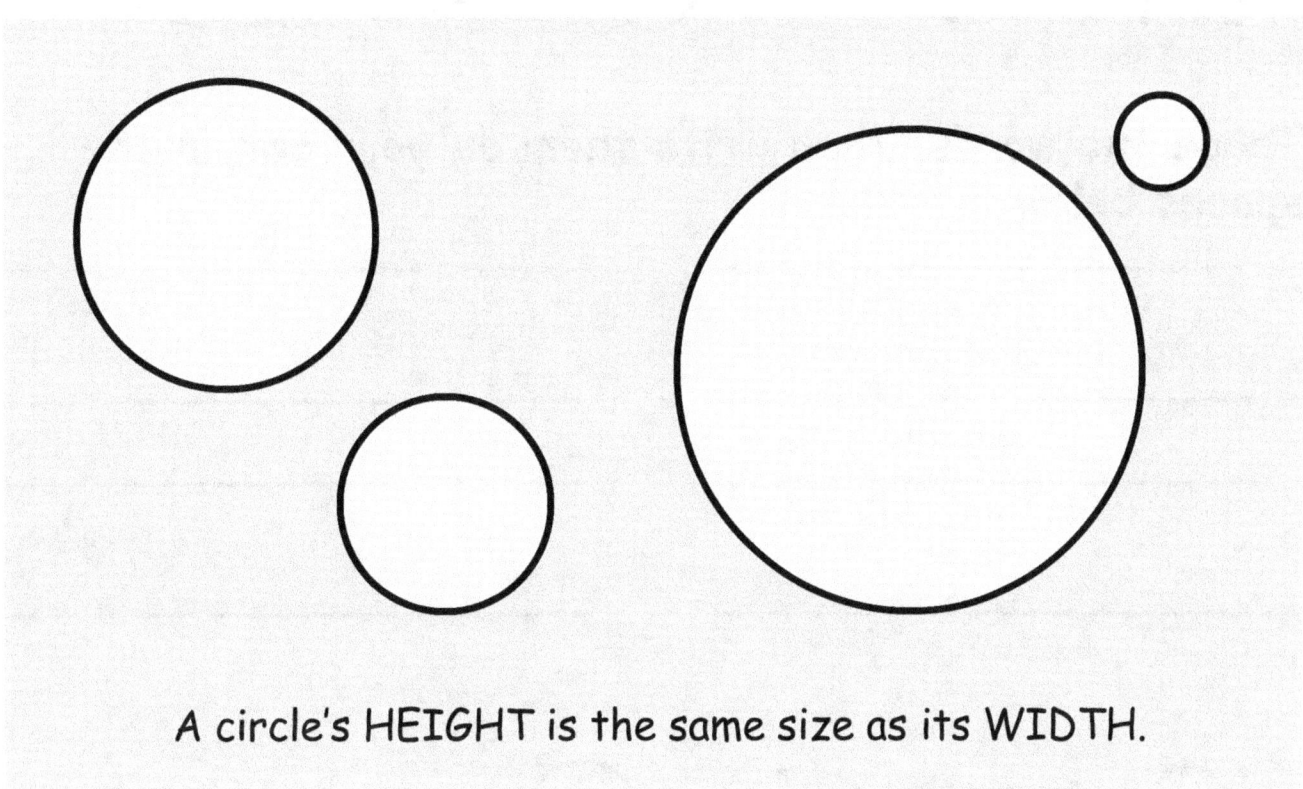

A circle's HEIGHT is the same size as its WIDTH.

Trace the circles below.

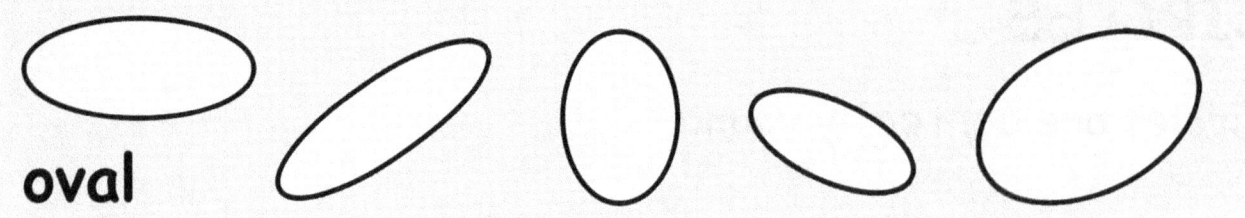

oval

Trace the words, then write them on your own in the spaces below:

oval

oblong

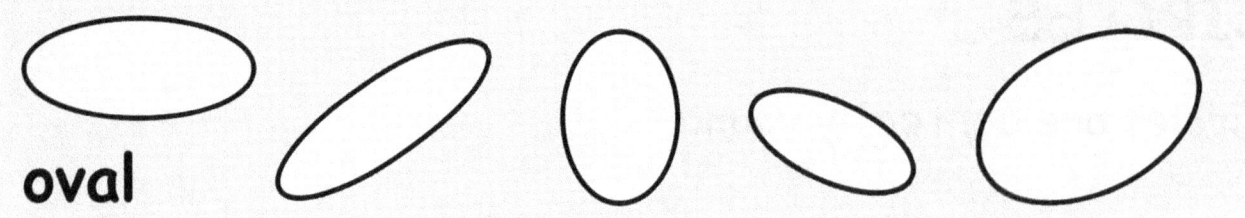

circle

Trace the words, then write them on your own in the spaces below:

circle

round

VERTICES

The **vertex** of a shape is like a corner. It is a point where 2 of its sides meet.

If there is more than one vertex, we call them **vertices**.

This shape has 4 vertices:

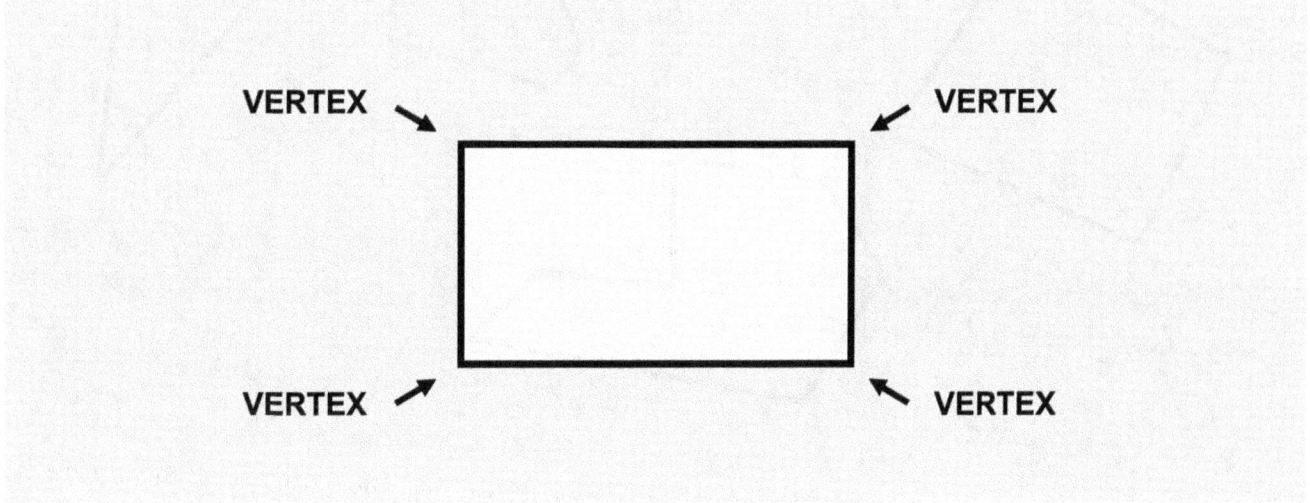

This shape has 3 vertices.

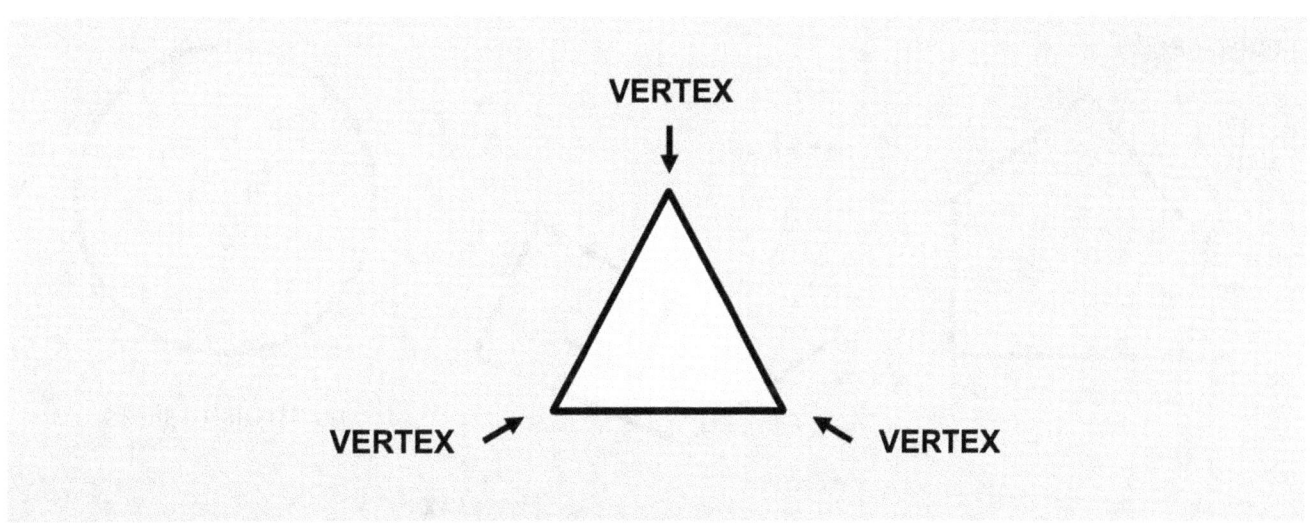

POLYGONS

Polygons are **closed** 2 Dimensional shapes that have **straight sides** and 3 or more **vertices**.

These are polygons:

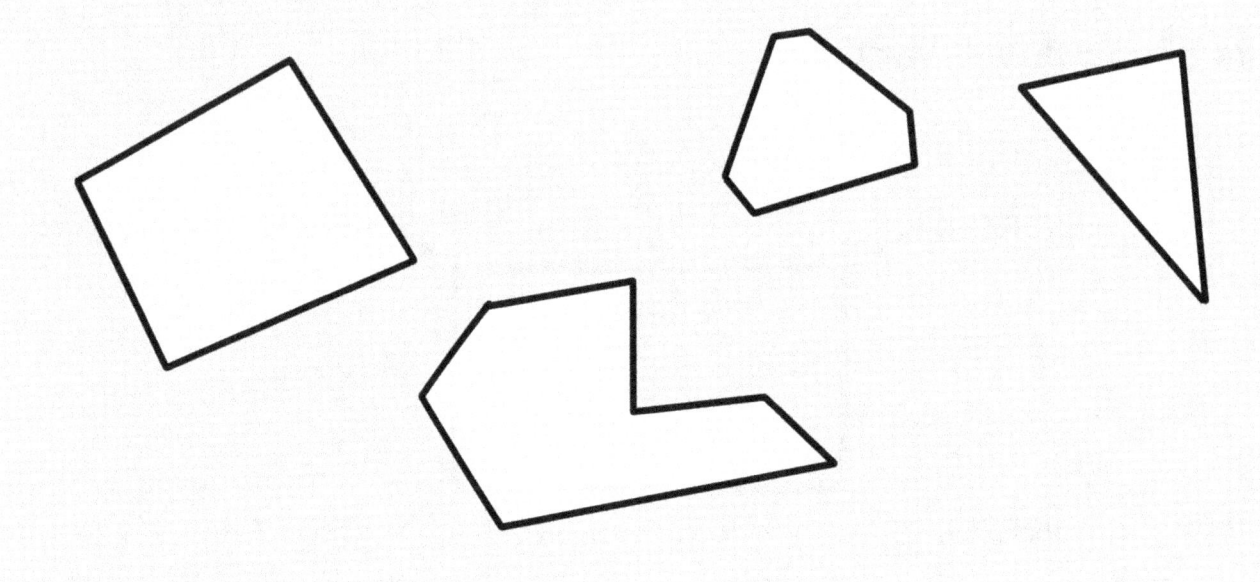

These are not polygons:

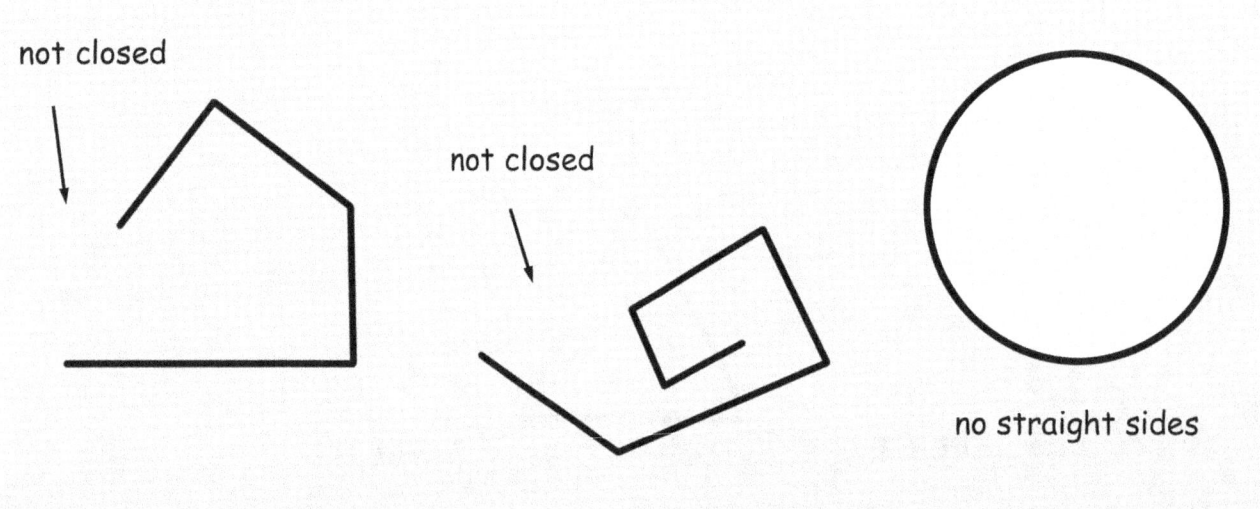

VERTICES

The **vertex** of a shape is like a corner. It is a point where 2 of its sides meet.

If there is more than one vertex, we call them **vertices**.

This shape has 4 vertices:

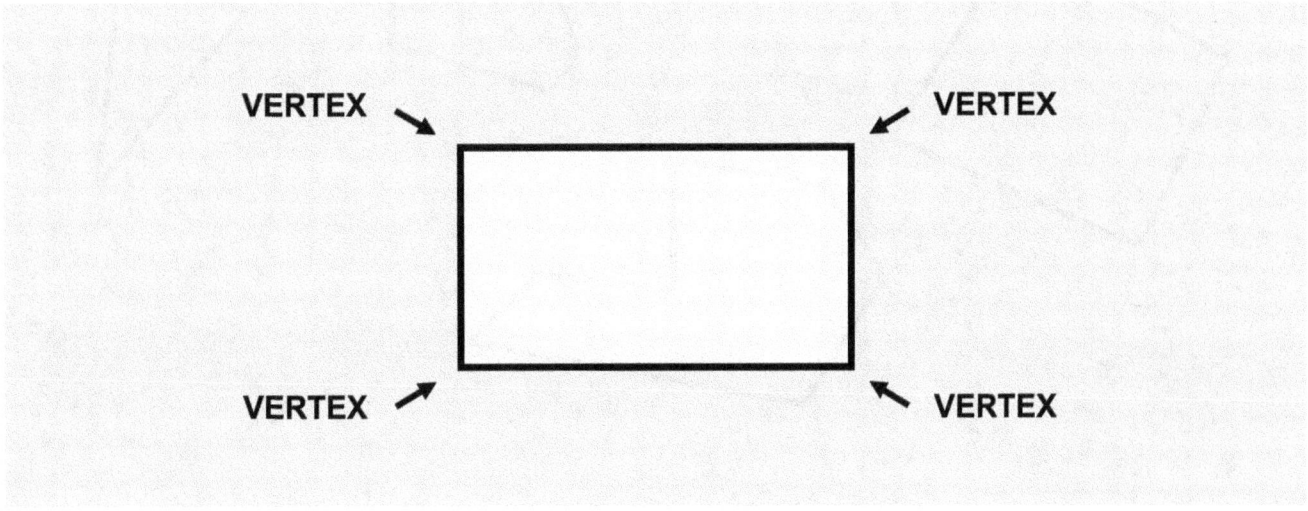

This shape has 3 vertices.

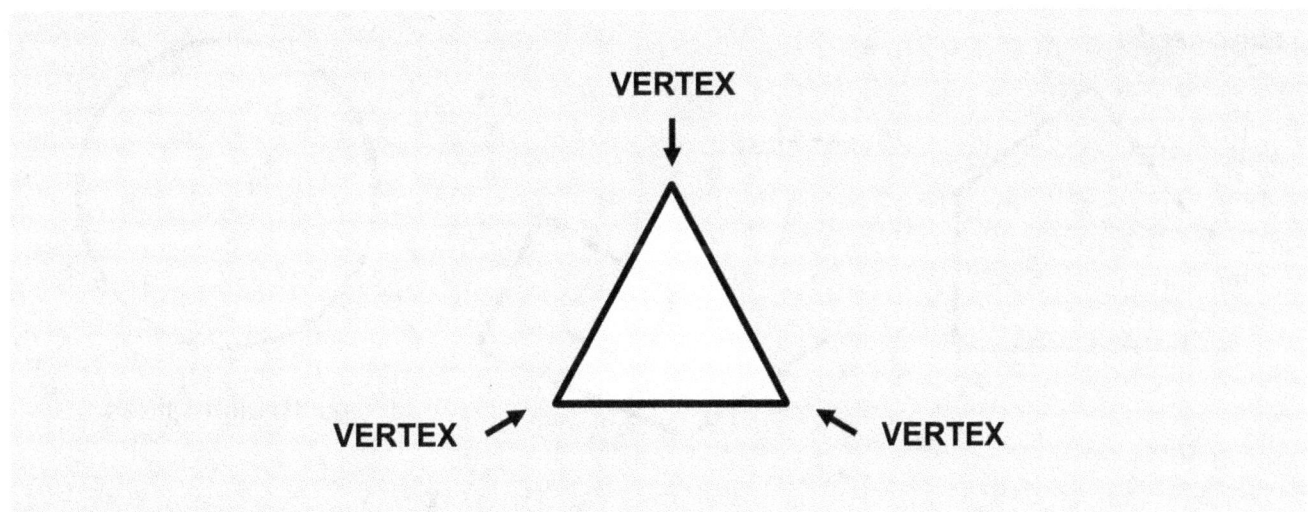

POLYGONS

Polygons are **closed** 2 Dimensional shapes that have **straight sides** and 3 or more **vertices**.

These are polygons:

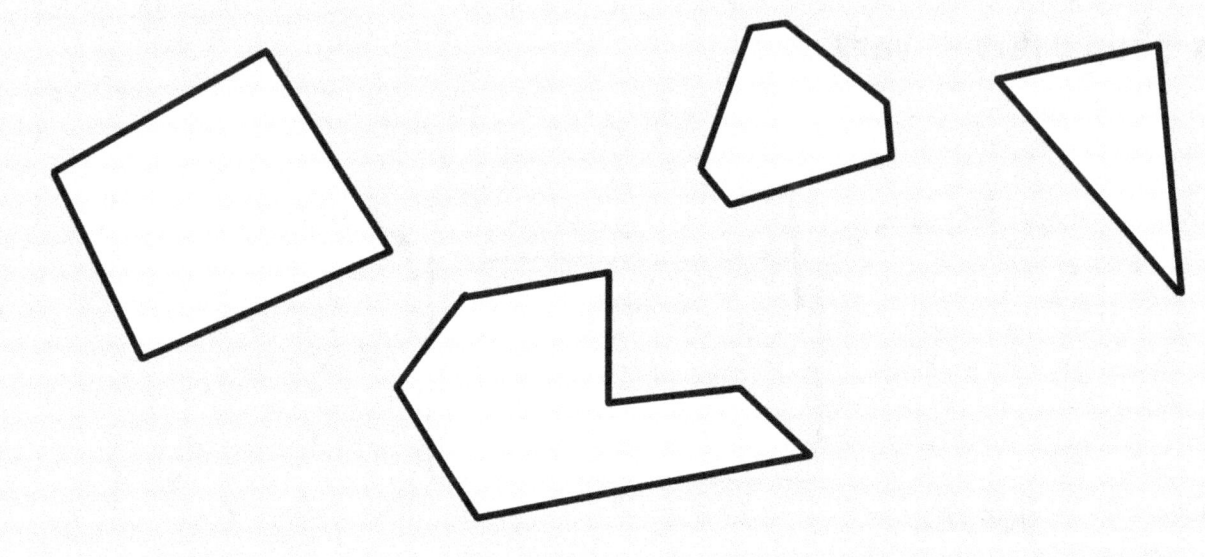

These are not polygons:

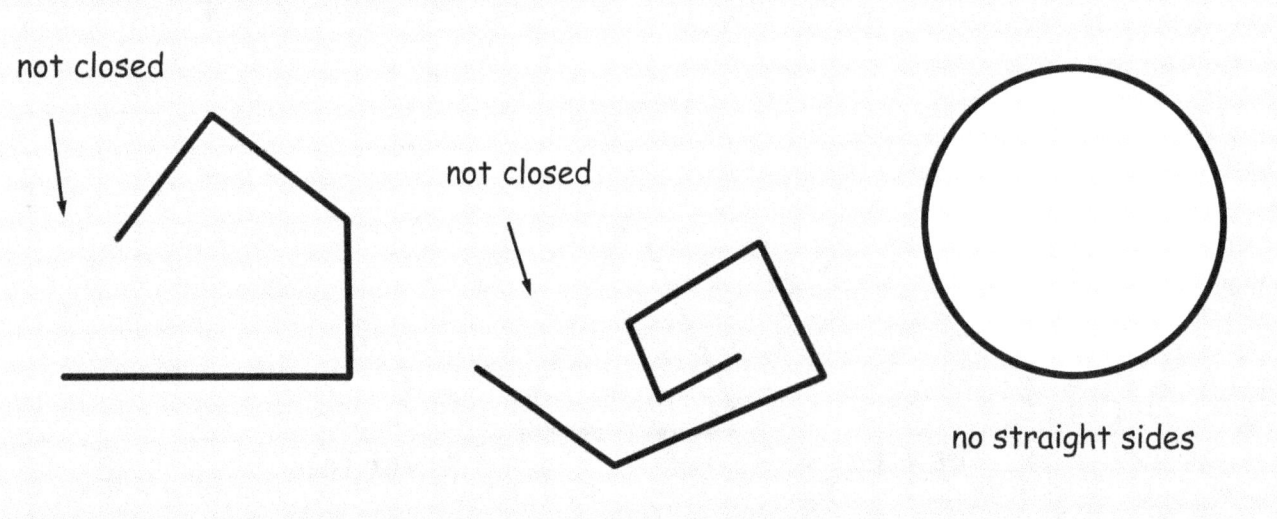

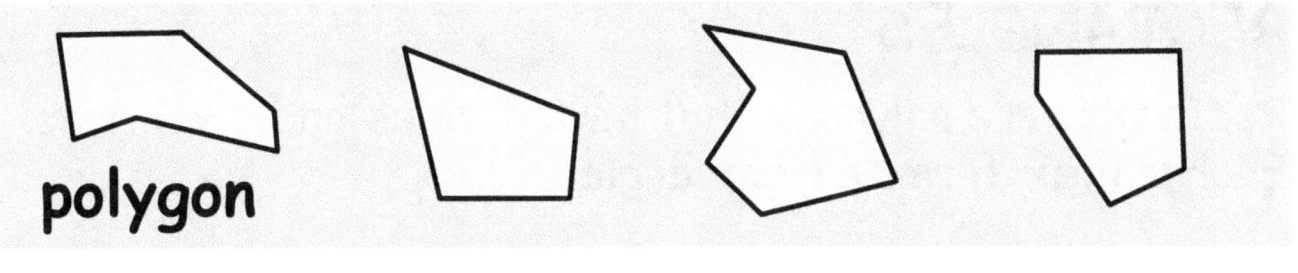

polygon

Trace the word, then write it on your own in the space below:

polygon

vertices

Trace the words, then write them on your own in the spaces below:

vertex   vertices

RECTANGLES

Rectangles are polygons that have **4 sides** and **4 vertices**. Each **vertex** forms a **right angle**.

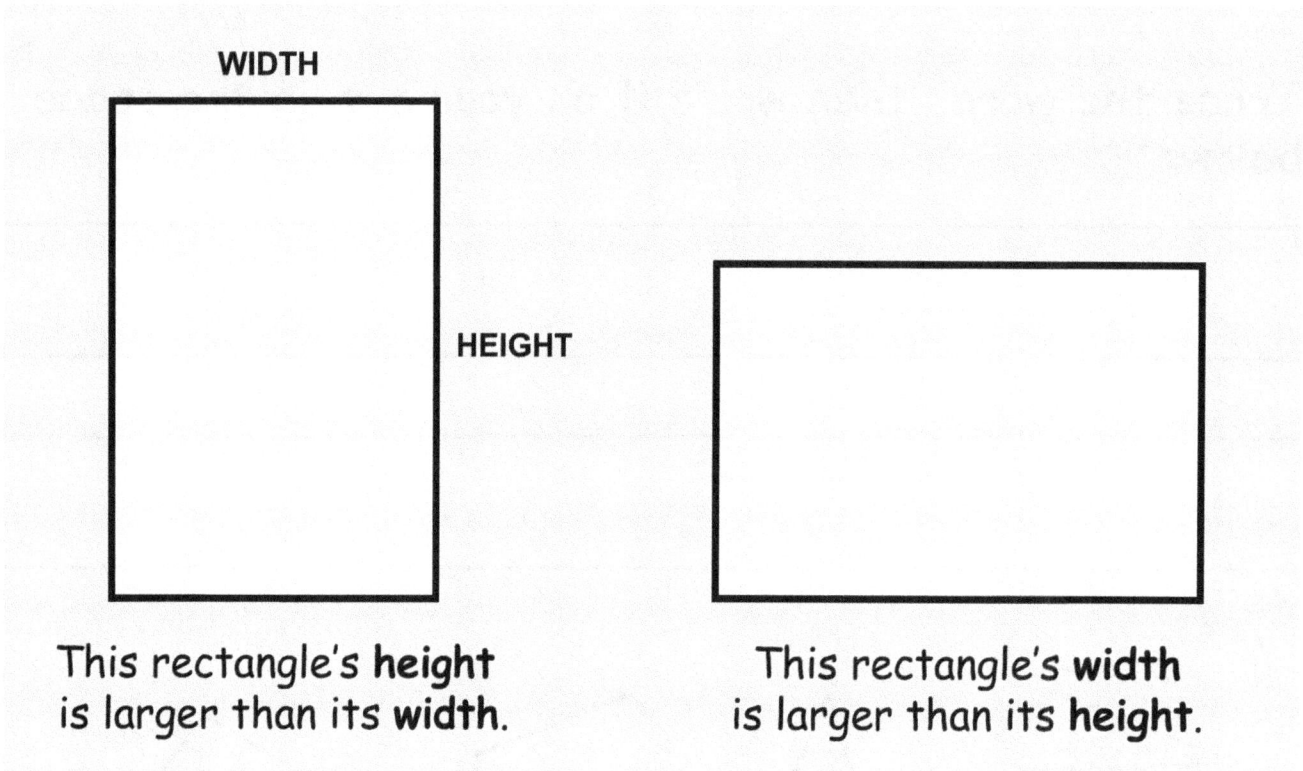

This rectangle's **height** is larger than its **width**.

This rectangle's **width** is larger than its **height**.

Trace the rectangles below.

SQUARES

Squares are polygons that have **4 sides** and **4 vertices**. Each **vertex** forms a **right angle** and **all sides of a square are the same size**.

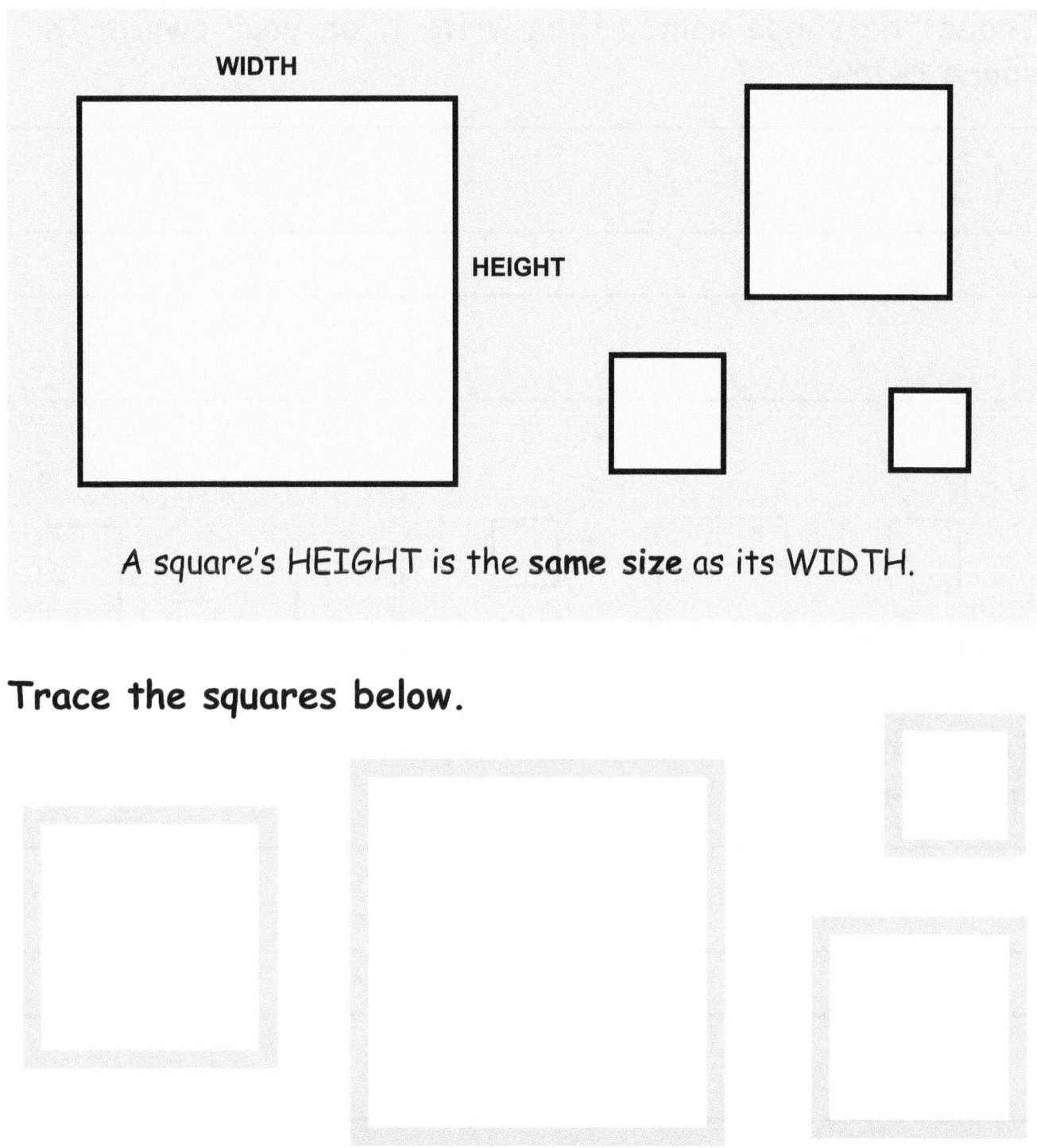

A square's HEIGHT is the **same size** as its WIDTH.

Trace the squares below.

rectangle

Trace the shape name, then write it on your own in the space below:

rectangle

square

Trace the shape name, then write it on your own in the space below:

square

TRIANGLES

Triangles are polygons that have **3 sides** and **3 vertices**.

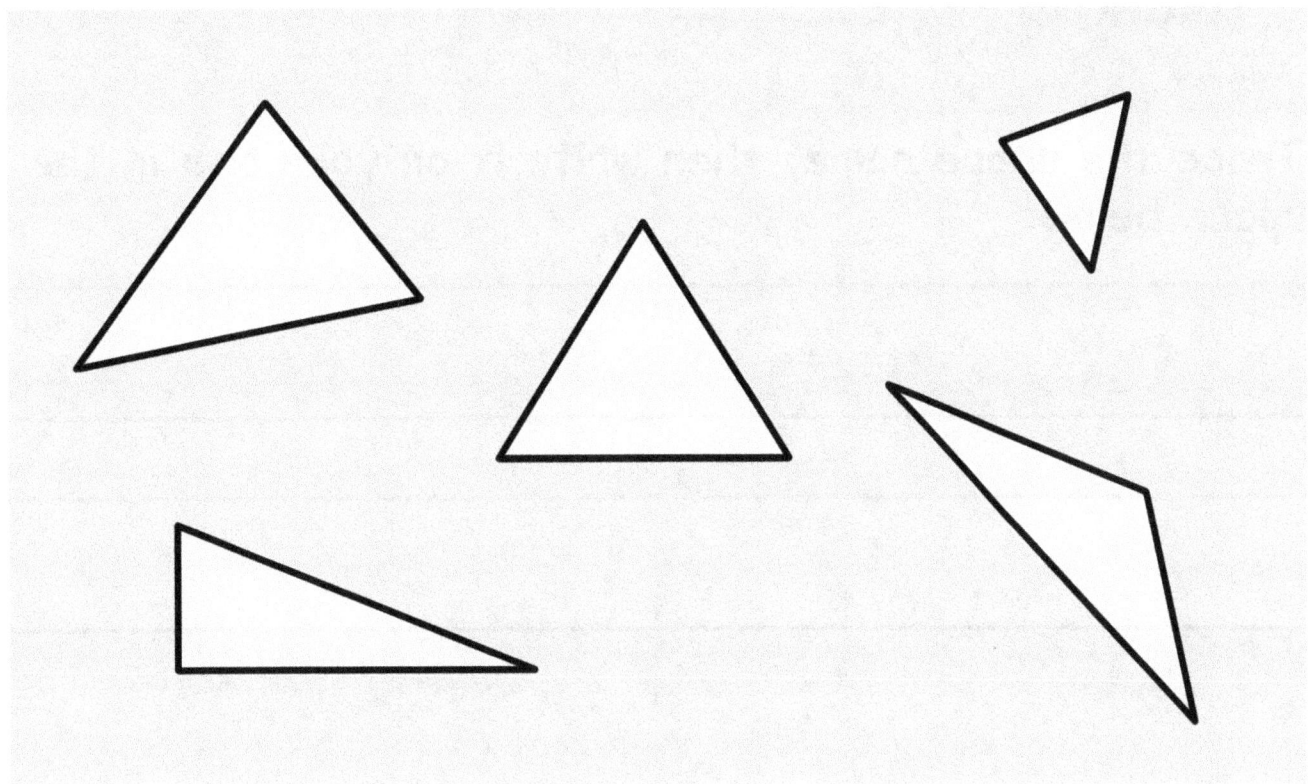

Trace the triangles below.

Trace the shape name, then write it on your own in the space below:

triangle

WORKSHEET
SHAPES & DIMENSIONS (OVALS & CIRCLES)

Trace the circles and ovals. Color the circles blue and then color the remaining ovals red.

How many circles are there? _____

WORKSHEET
SHAPES & DIMENSIONS (RECTANGLES & SQUARES)

Trace the squares and rectangles. Color the squares blue and then color the remaining rectangles red.

How many squares are there? _____

WORKSHEET
SHAPES & DIMENSIONS (WIDTH & HEIGHT)

Color the rectangles red whose width is larger than its height.

Color the rectangles blue whose height is larger than its width.

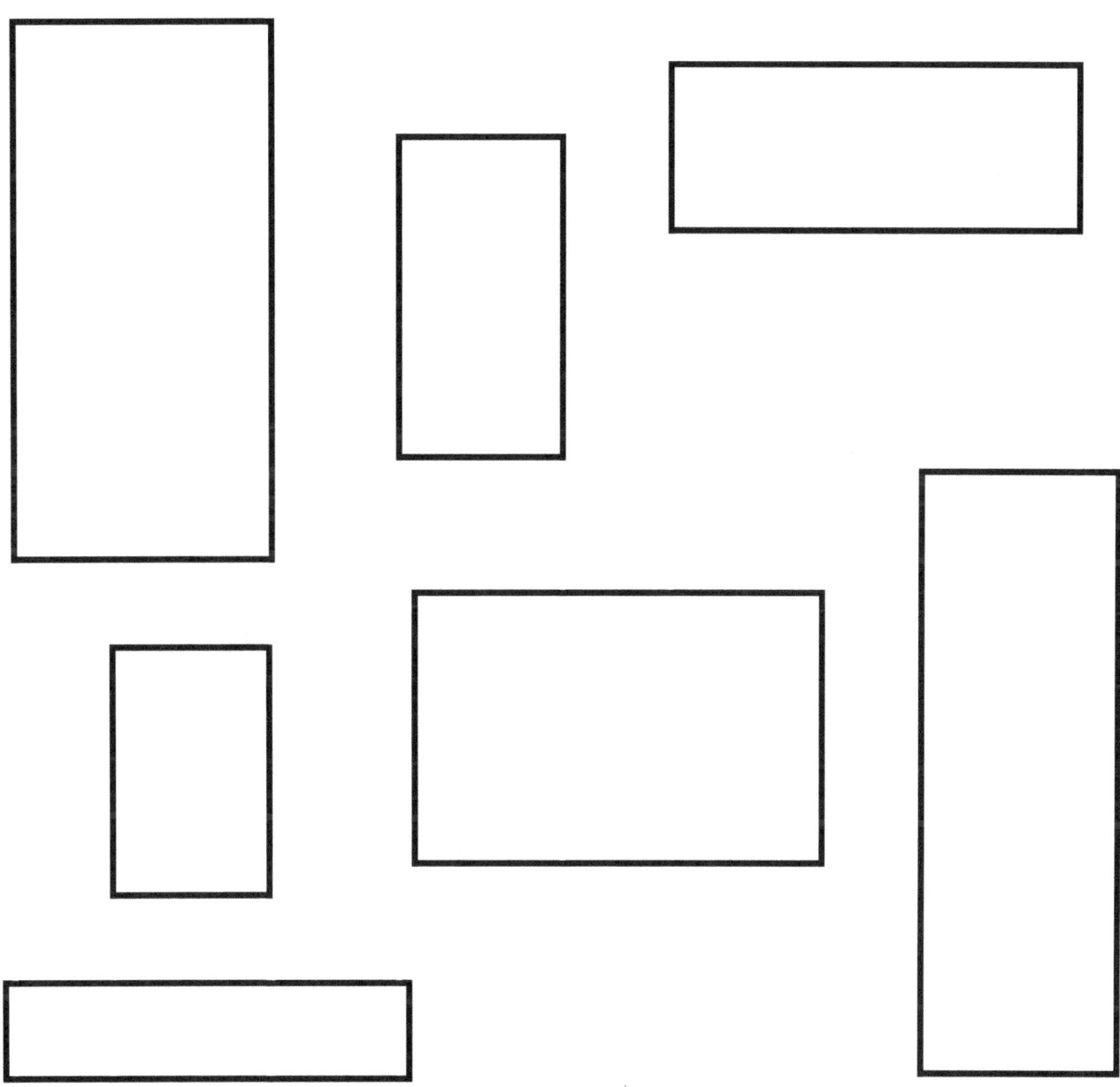

WORKSHEET
SHAPES & DIMENSIONS (WIDTH & HEIGHT)

Color the ovals red whose width is larger than its height.

Color the ovals blue whose height is larger than its width.

WORKSHEET
SHAPES & DIMENSIONS (VERTICES)

Count the number of vertices in each of the shapes and write the amounts in the blanks.

1.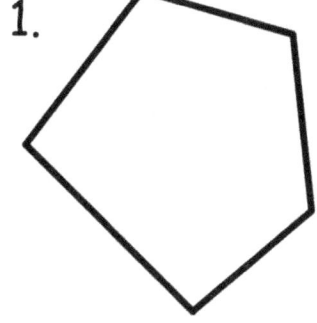
_____

2.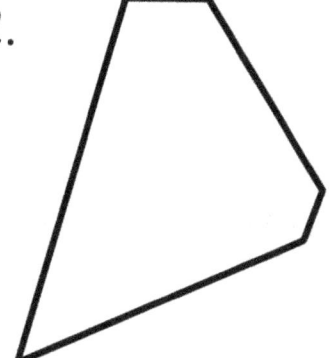
_____

3.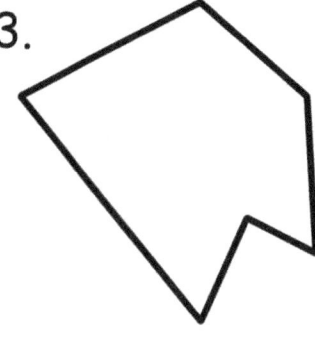
_____

4.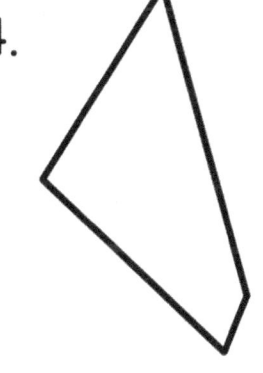
_____

5.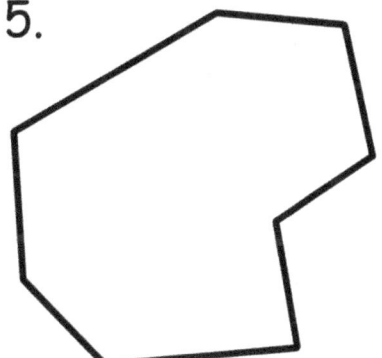
_____

6.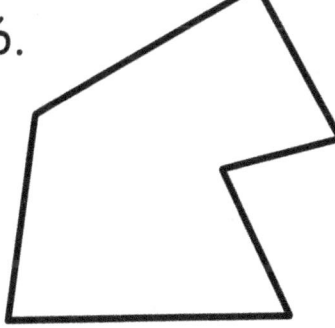
_____

WORKSHEET
SHAPES & DIMENSIONS (2-DIMENSIONAL SHAPES)

Count the different types of 2 dimensional shapes, then write the correct amounts in the blanks.

CIRCLES _____   SQUARES _____

OVALS _____   RECTANGLES _____

TRIANGLES _____

Lesson 7: Shapes & Dimensions

LESSON 8

Regular Polygons

REGULAR POLYGONS

A **regular polygon** is a polygon whose sides are all the same length and whose angles are all the same size.

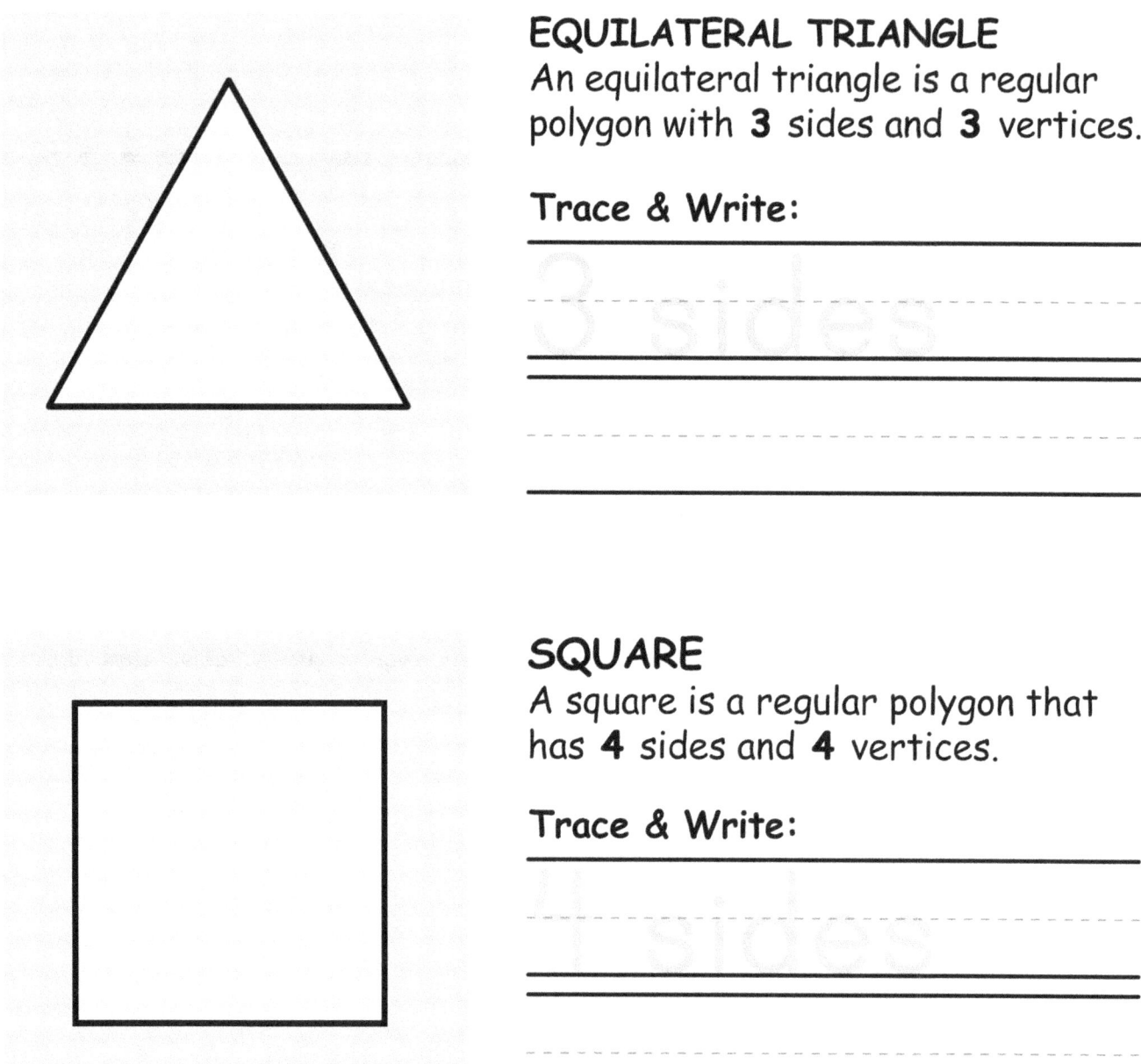

EQUILATERAL TRIANGLE
An equilateral triangle is a regular polygon with **3** sides and **3** vertices.

Trace & Write:

3 sides

SQUARE
A square is a regular polygon that has **4** sides and **4** vertices.

Trace & Write:

4 sides

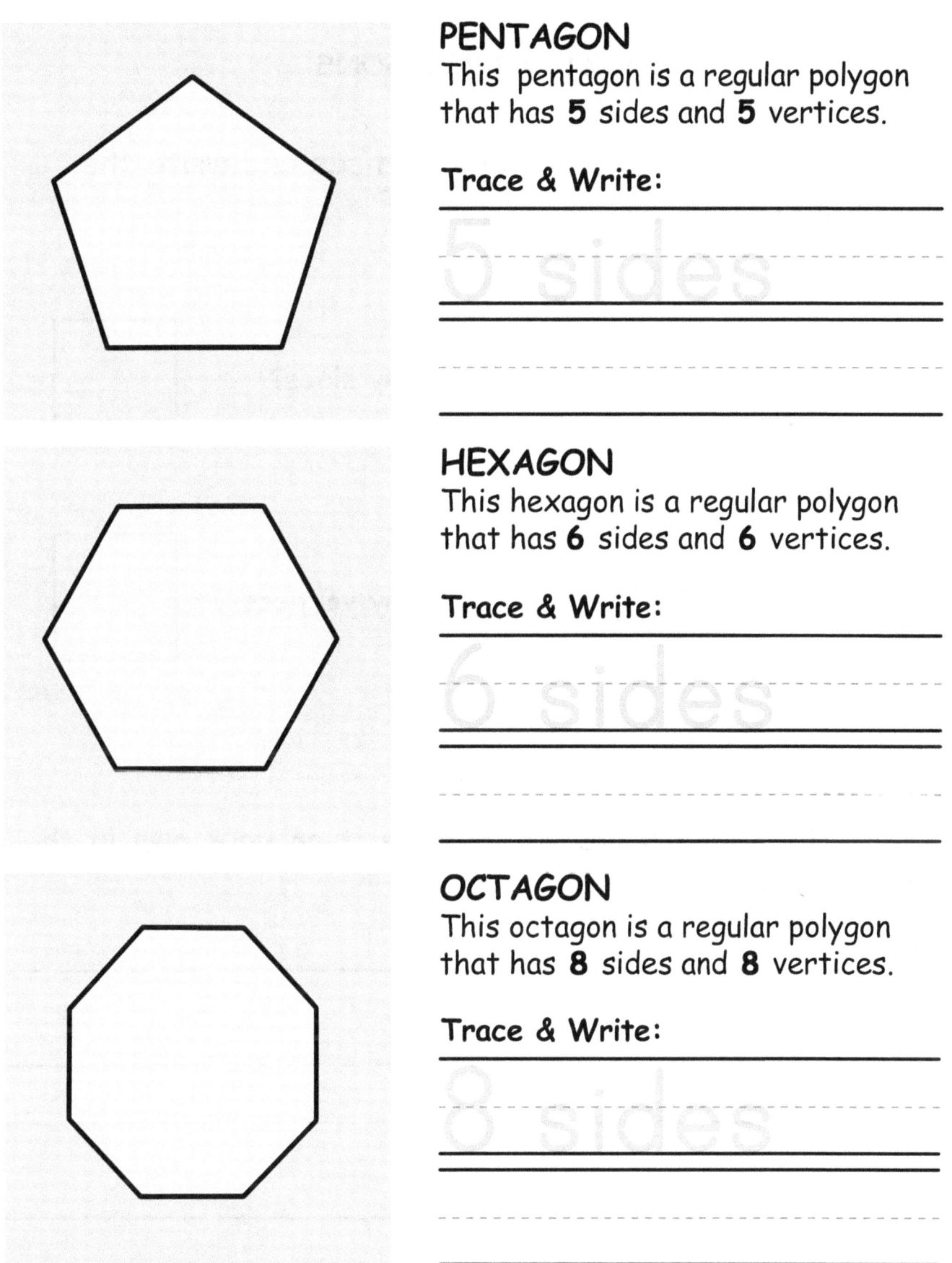

PENTAGON

This pentagon is a regular polygon that has **5** sides and **5** vertices.

Trace & Write:

5 sides

HEXAGON

This hexagon is a regular polygon that has **6** sides and **6** vertices.

Trace & Write:

6 sides

OCTAGON

This octagon is a regular polygon that has **8** sides and **8** vertices.

Trace & Write:

8 sides

WORKSHEET
REGULAR POLYGONS

Count the number of sides and vertices and write the amounts in the blanks.

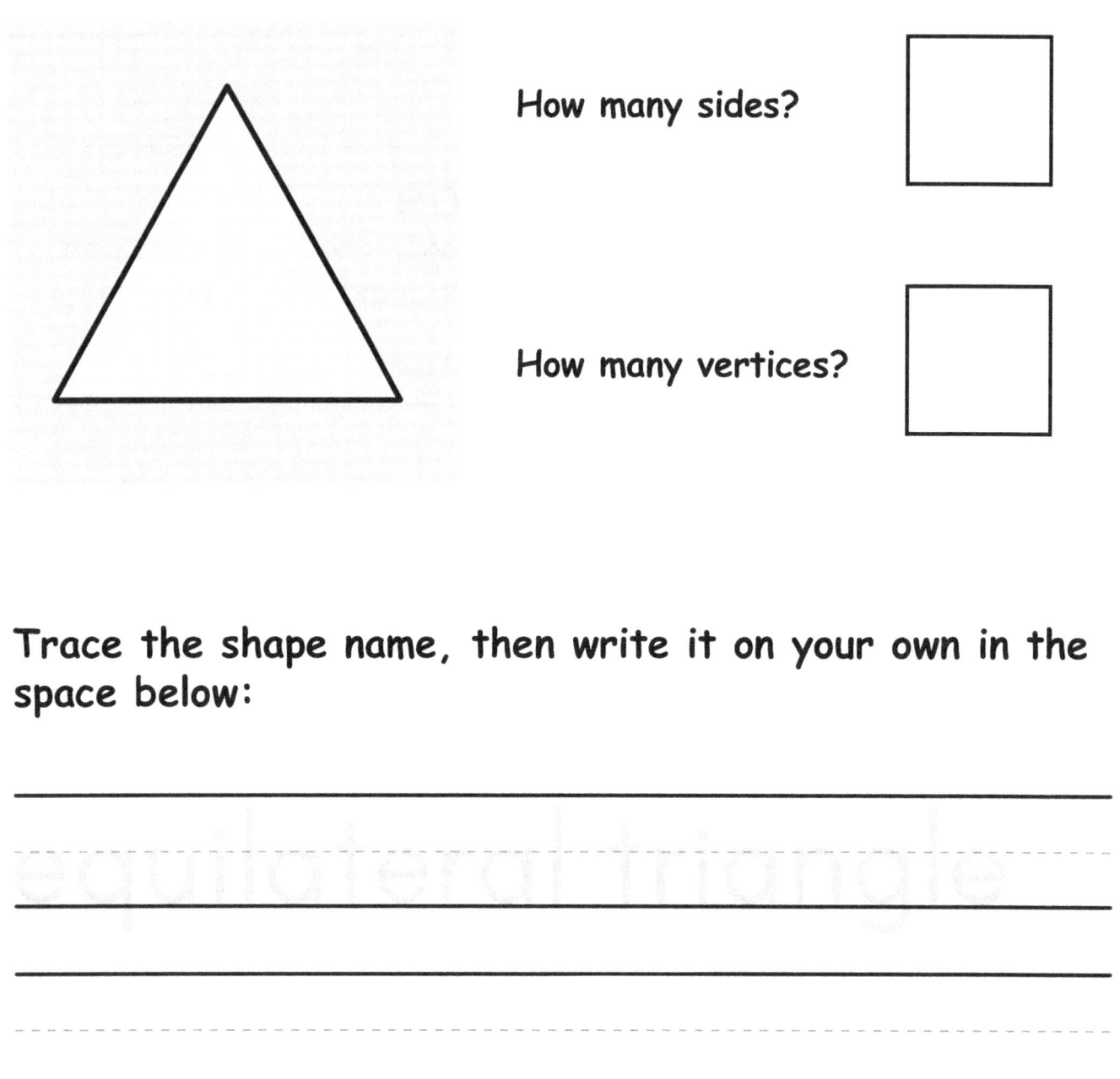

How many sides?

How many vertices?

Trace the shape name, then write it on your own in the space below:

_____

equilateral triangle

_____

_____

WORKSHEET
REGULAR POLYGONS

Count the number of sides and vertices and write the amounts in the blanks.

How many sides?

How many vertices?

Trace the shape name, then write it on your own in the space below:

_____

square

_____

_____

WORKSHEET
REGULAR POLYGONS

Count the number of sides and vertices and write the amounts in the blanks.

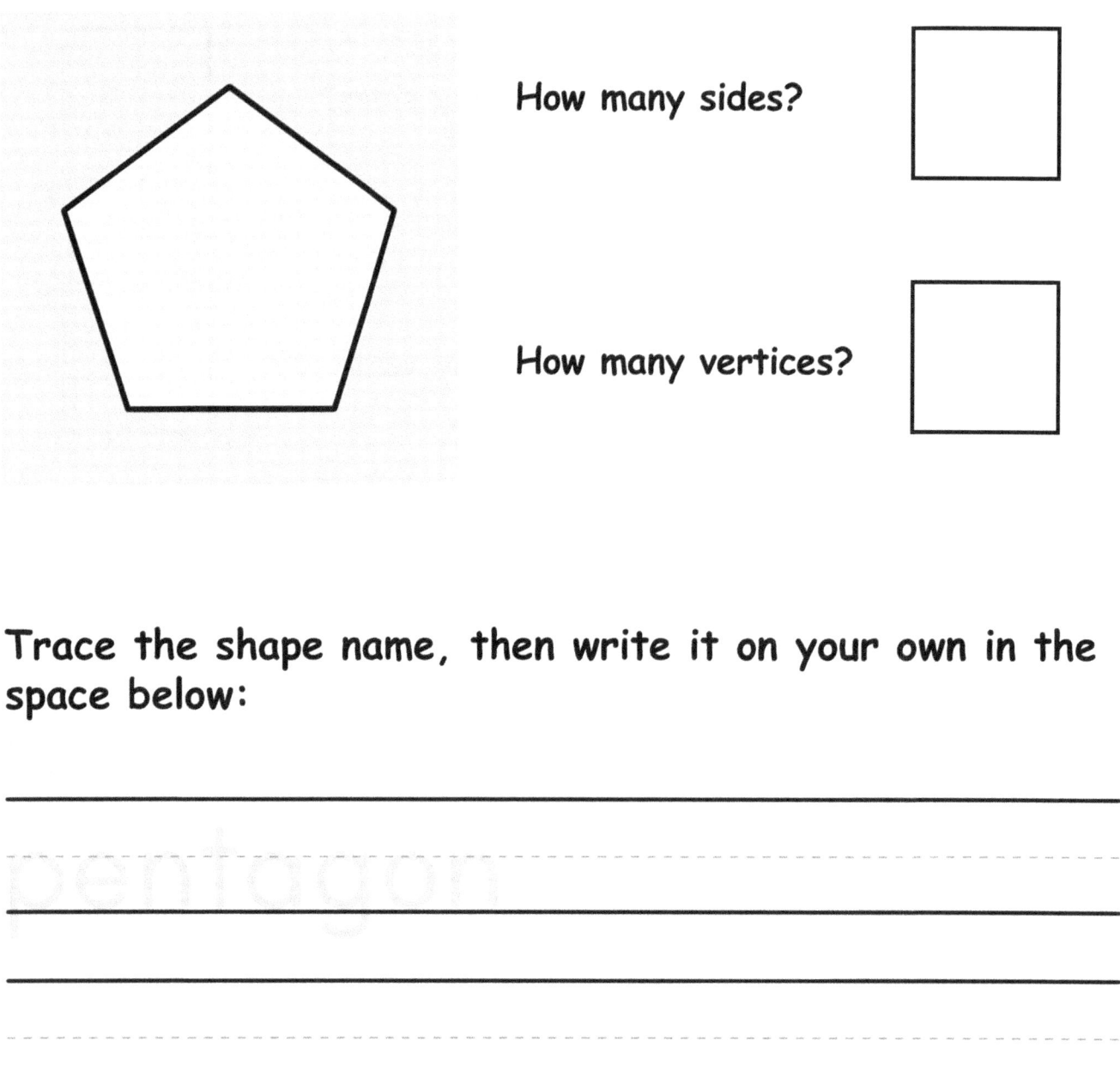

How many sides? ☐

How many vertices? ☐

Trace the shape name, then write it on your own in the space below:

_____

pentagon

_____

_____

WORKSHEET
REGULAR POLYGONS

Count the number of sides and vertices and write the amounts in the blanks.

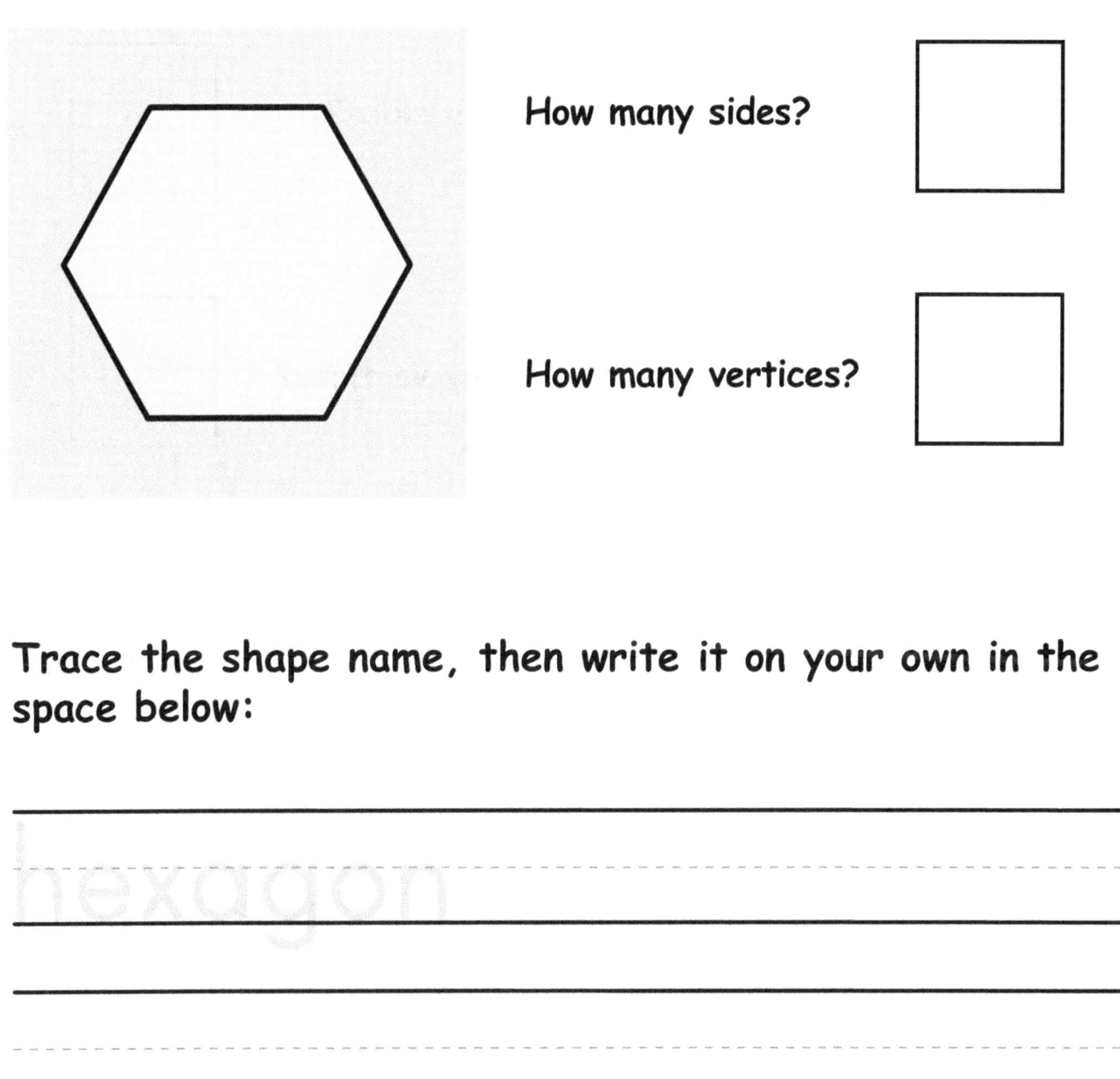

How many sides?

How many vertices?

Trace the shape name, then write it on your own in the space below:

hexagon

WORKSHEET
REGULAR POLYGONS

Count the number of sides and vertices and write the amounts in the blanks.

How many sides?

How many vertices?

Trace the shape name, then write it on your own in the space below:

octagon

WORKSHEET
REGULAR POLYGONS

Draw a line from each shape to its name.

HEXAGON

SQUARE

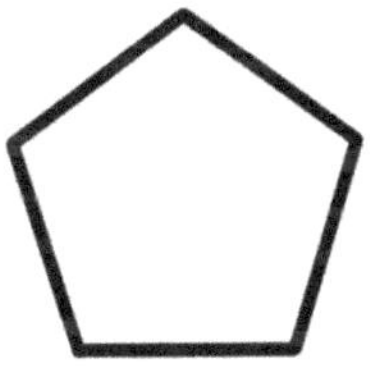

OCTAGON

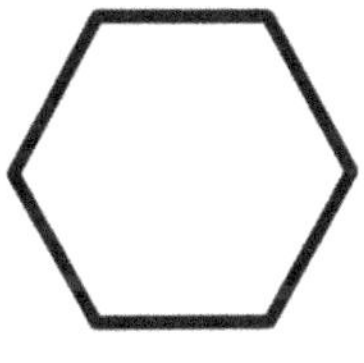

EQUILATERAL TRIANGLE

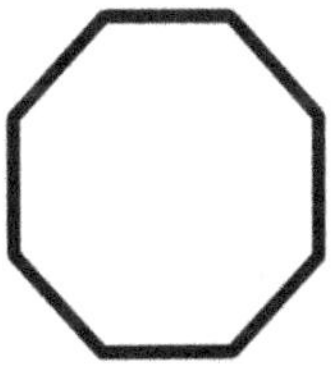

PENTAGON

Lesson 8: Regular Polygons

LESSON 9

3-Dimensional Shapes

3-DIMENSIONAL SHAPES

2-dimensional shapes (or 2D shapes) like circles, squares and triangles are **flat**.

3-dimensional shapes (or 3D shapes) are **not flat**. Things like balls and blocks are 3-dimensional.

2D shapes have **2** dimensions: **width & height**.
3D shapes have **3** dimensions: **width, height & depth**.

SQUARE
A square is flat.
It has 2 dimensions:
WIDTH & HEIGHT

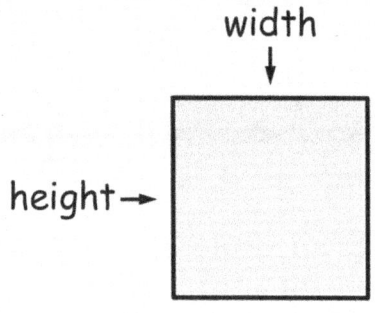

CUBE
A cube is not flat.
It has 3 dimensions:
WIDTH, HEIGHT & DEPTH

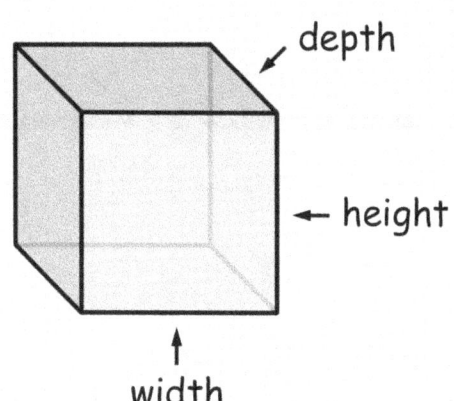

The outside part of a 3D shape is called the **surface**.

Some 3D shapes, like cubes, have flat 2D **surfaces** called **faces** and some 3d shapes have curved **surfaces** ... like balls.

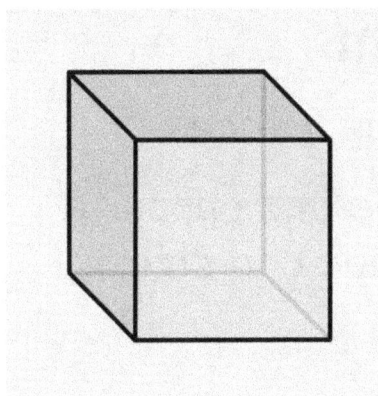

Cube

A **cube** has 6 faces.

Each face is a **square** and each face is the same size.

Trace the shape name, then write it on your own in the space below:

cube

Draw a Cube
Trace the light grey lines in each step.

 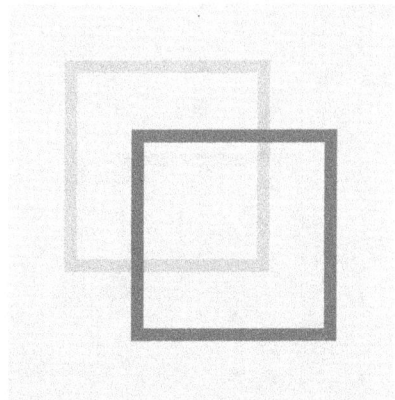 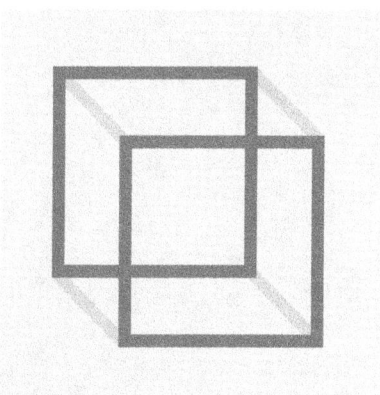

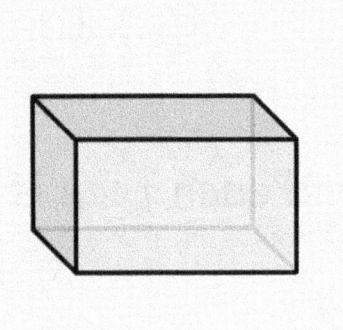

Rectangular Prism

A **rectangular prism** has 6 faces.

Each face is a **rectangle** and opposite faces are the same shape and size.

Trace the shape name, then write it on your own in the space below:

rectangular prism

Draw a Rectangular Prism
Trace the light grey lines in each step.

Triangular Prism

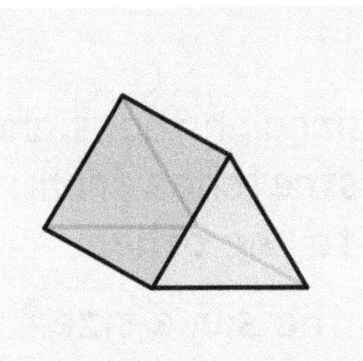

A **triangular prism** has 5 faces.

The 2 opposite faces are **triangles** and the other 3 faces are **rectangles**.

Trace the shape name, then write it on your own in the space below:

triangular prism

Draw a Triangular Prism
Trace the light grey lines in each step.

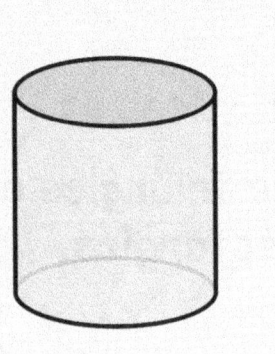

Cylinder

A **cylinder** has 2 flat, circular faces and a curved surface that stretches from the edge of one circle to the other.

The 2 **circles** are both the same size.

Trace the shape name, then write it on your own in the space below:

cylinder

Draw a Cylinder

Trace the light grey lines in each step.

The ends of a cylinder are round like a circle, but when you draw one, draw the ends as ovals. This makes it look like you are seeing it from the side and gives it a 3D look.

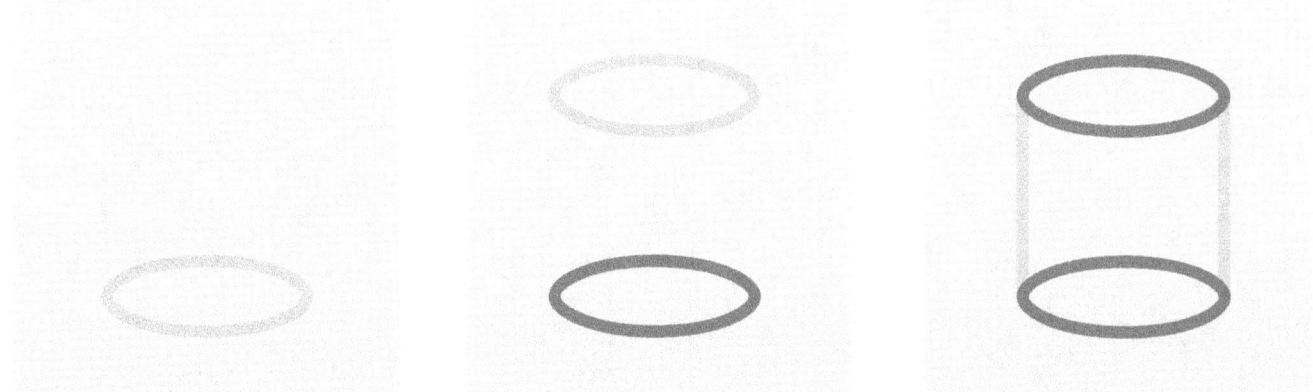

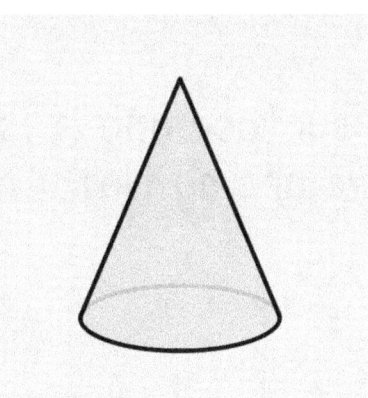

Cone

A **cone** has 1 flat, circular face and a curved surface that stretches from the edge of the circle to a point above it.

Trace the shape name, then write it on your own in the space below:

----
cone

Draw a Cone

Trace the light grey lines in each step.

The second step has just a dot above the base of the cone to mark where the 2 lines will meet.

Lesson 9: 3D Shapes

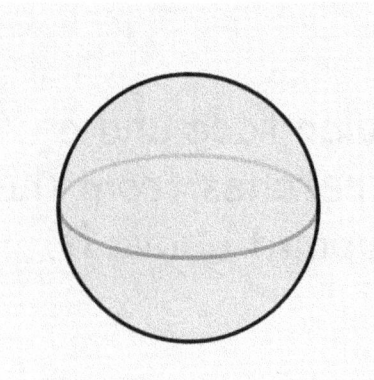

Sphere

A **sphere** has a curved surface and it is shaped like a ball. There are no flat faces.

Trace the shape name, then write it on your own in the space below:

sphere

Draw a Sphere

Trace the light grey lines in each step.

The oval is drawn in the middle of the circle to give it a 3D look.

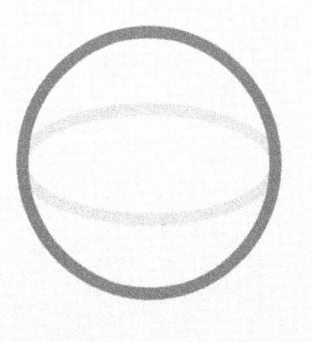

Trace the words, then write them on your own in the spaces below:

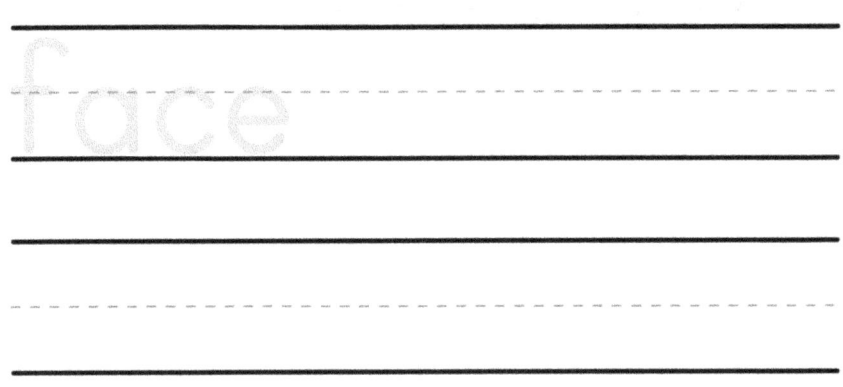
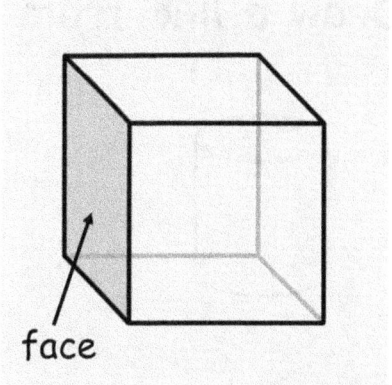
face

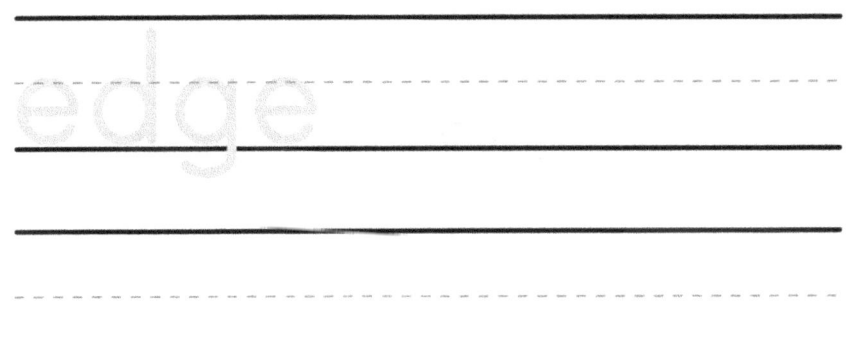
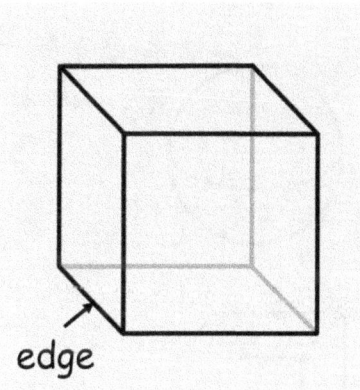
edge

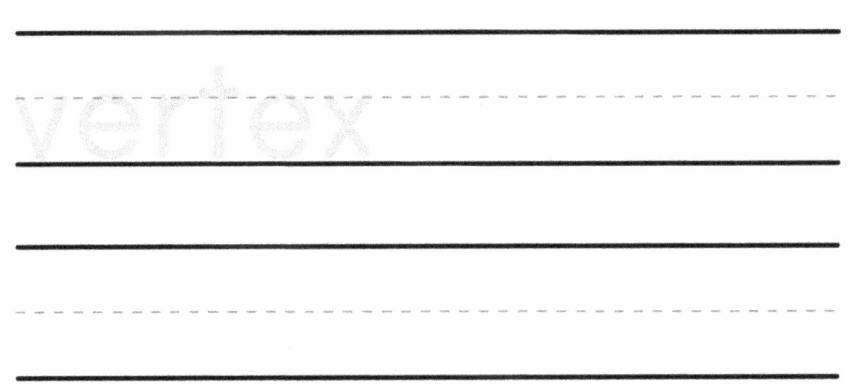
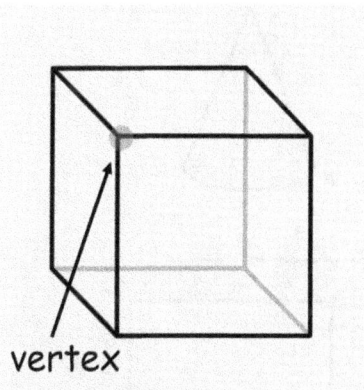
vertex

WORKSHEET
3-DIMENSIONAL (3D) SHAPES

Draw a line from each 3D shape to its name.

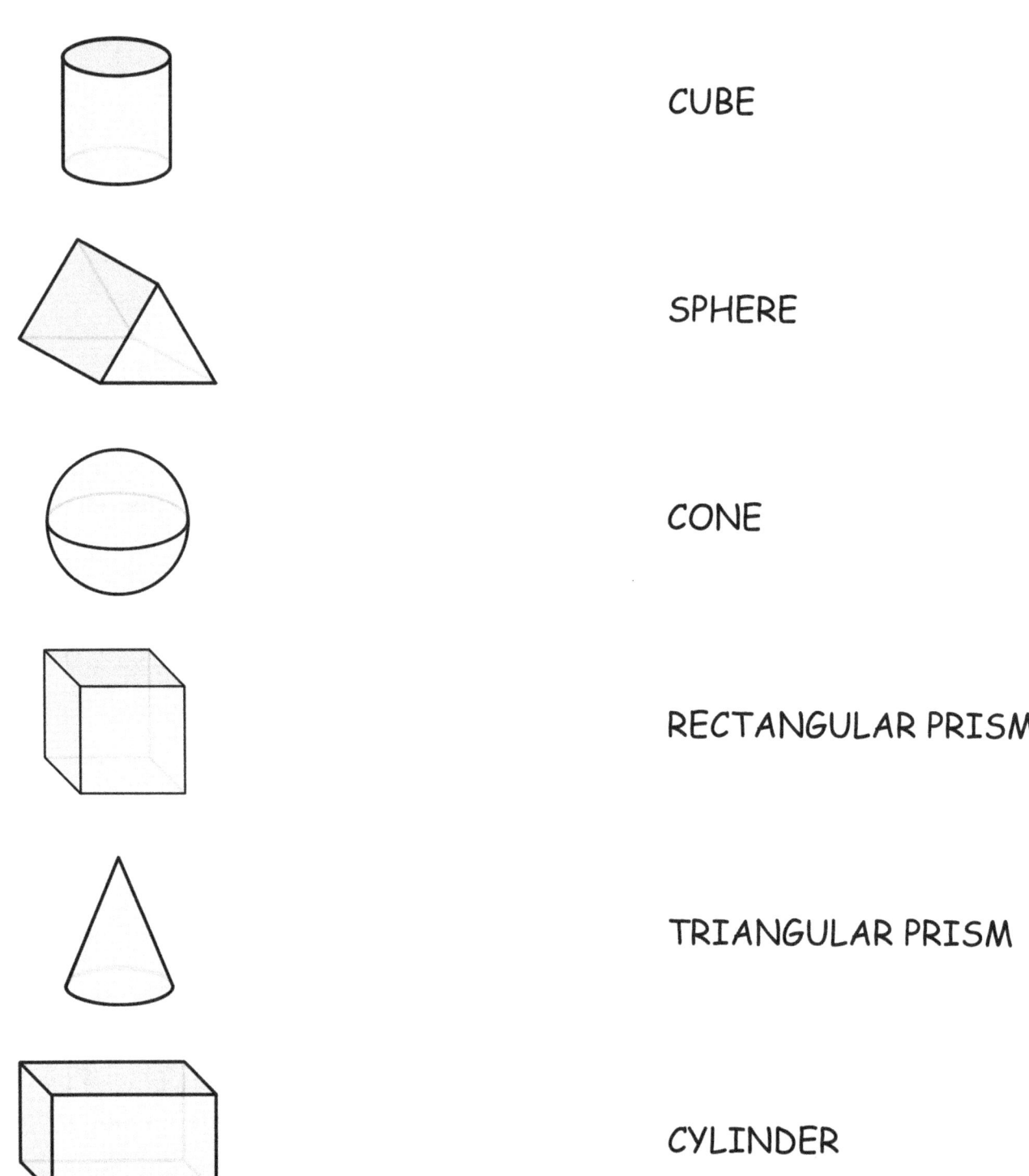

WORKSHEET
3-DIMENSIONAL (3D) SHAPES

Write the name of the shape in the lines below it and draw the shape in the blank space next to it.

Cube

_____

Rectangular Prism

_____

WORKSHEET
3-DIMENSIONAL (3D) SHAPES

Write the name of the shape in the lines below it and draw the shape in the blank space next to it.

Triangular Prism

_____

Cylinder

_____

Lesson 9: 3D Shapes

WORKSHEET
3-DIMENSIONAL (3D) SHAPES

Write the name of the shape in the lines below it and draw the shape in the blank space next to it.

Cone

Sphere

WORKSHEET
3-DIMENSIONAL (3D) SHAPES

Count the different 3D shapes and write the correct amounts in the blanks.

CUBES _____   CYLINDERS _____

RECTANGULAR PRISMS _____   CONES _____

TRIANGULAR PRISMS _____   SPHERES _____

LESSON 10

Addition

ADDITION

Addition is used to find the total amount of items we combine together. That total amount is called the **sum**.

Suppose you have 2 apples and your friend has 1 apple. We use **addition** to find out how many apples you and your friend have **all together**.

One way to find the sum of these apples is to just count each apple:

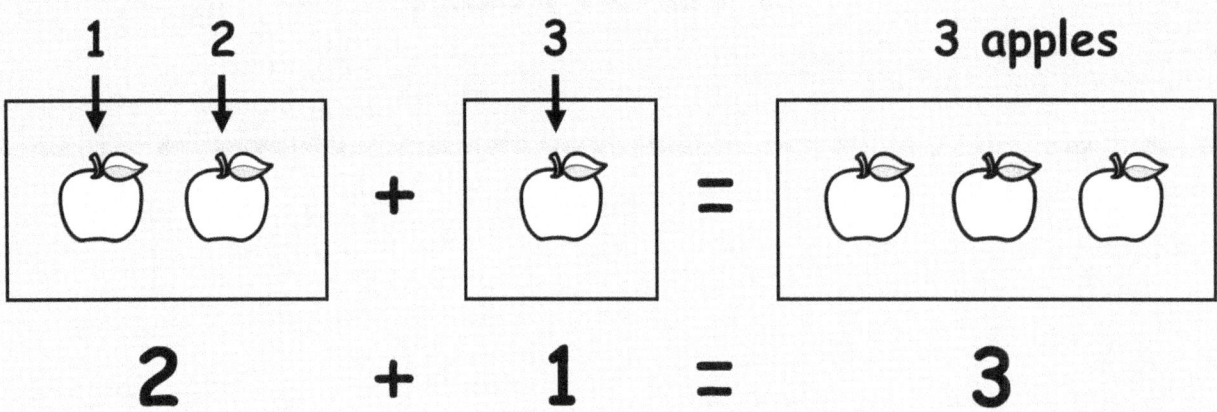

The **plus sign** "+" is used to show we are **adding** apples together. After the addition is the **equal sign** "=" and then the **sum**.

By counting, you can see there are 3 apples all together.

Example: Find the sum: 3 + 2

Let's use the pictures to figure this out by counting 3 + 2 balls.

First: We color in 3 balls, then color in 2 more. That's **3 + 2** balls.

Next: We count the total number of balls that are colored in to find the sum.

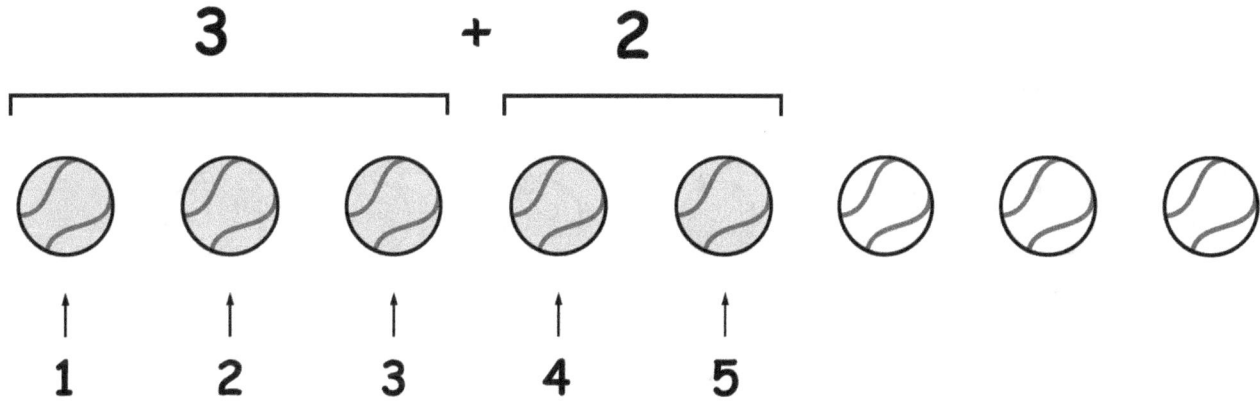

There are 5 balls colored in all together.

Trace the statement, then write it on your own in the space below:

3 + 2 = 5

Example: Find the sum: 2 + 2

Let's use the pictures to figure this out by counting 2+2 oranges.

First: We color in 2 oranges, then color in 2 more.
That's **2 + 2** oranges.

Next: We count the total number of oranges that are colored in to find the sum.

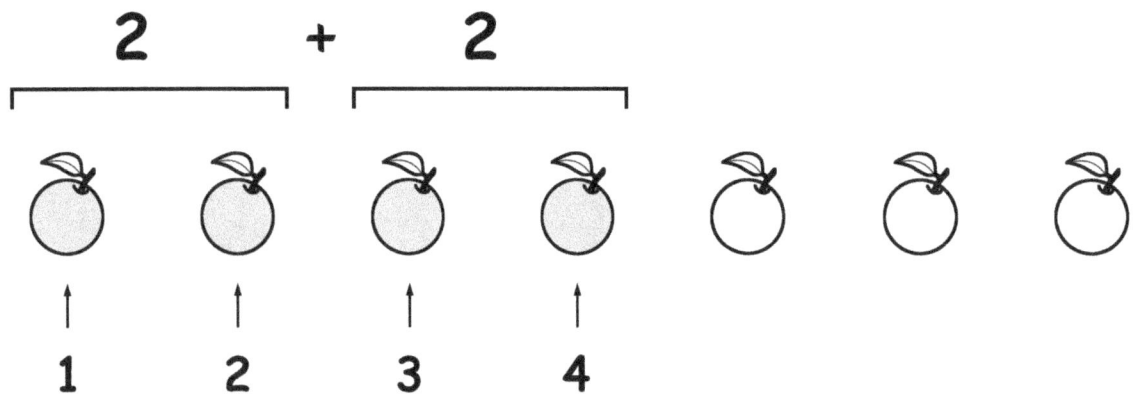

There are 4 oranges colored in all together.

Trace the statement, then write it on your own in the space below:

2 + 2 = 4

Example: Find the sum: 4 + 2

Let's use the pictures to figure this out by counting 4 + 2 fish.

First: We color in 4 fish, then color in 2 more.
That's **4 + 2** fish.

Next: We count the total number of fish that are colored in to find the sum.

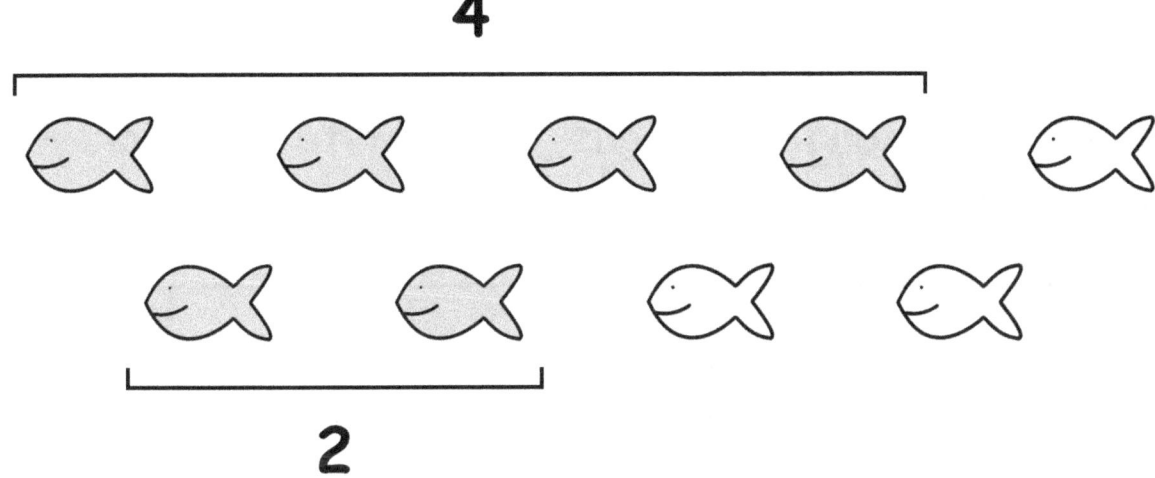

There are 6 fish colored in all together.

Trace the statement, then write it on your own in the space below:

4 + 2 = 6

Lesson 10: Addition

Trace the word, then write it on your own in the space below:

addition

Trace the word and plus signs, then write them on your own in the space below:

plus   +   +   +

Trace the word, then write it on your own in the space below:

sum

WORKSHEET
ADDITION

1. Find the sum: 6 + 1

Trace and complete the addition statement below.

Color in 6 smilies, then color in 1 more smilie to help find the sum.

6 + 1 =

2. Find the sum: 5 + 2

Trace and complete the addition statement below.

Color in 5 beach balls, then color in 2 more beach balls to help find the sum.

5 + 2 =

WORKSHEET
ADDITION

1. Find the sum: 8 + 4

Trace and complete the addition statement below.

Color in 8 apples, then color in 4 more apples to help find the sum.

8 + 4 =

2. Find the sum: 6 + 5

Trace and complete the addition statement below.

Color in 6 stars, then color in 5 more stars to help find the sum.

6 + 5 =

WORKSHEET
ADDITION

1. Find the sum: 4 + 4

Trace and complete the addition statement below.

Color in 4 jelly beans, then color in 4 more jelly beans to help find the sum.

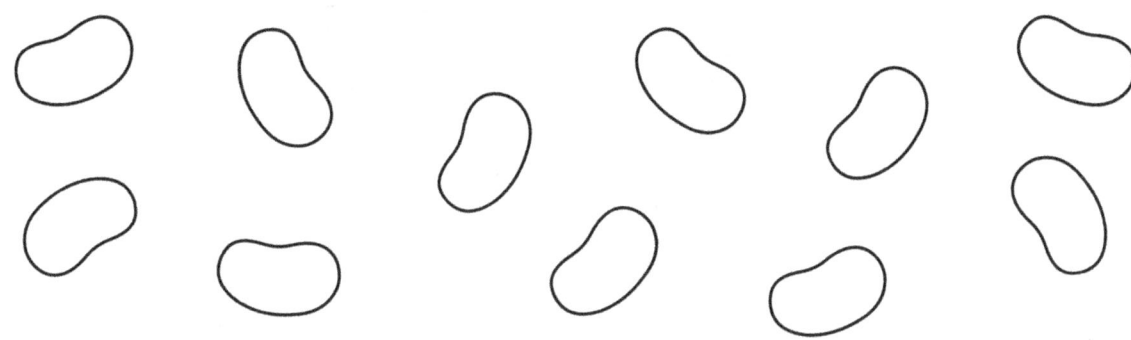

4 + 4 =

2. Find the sum: 7 + 5

Trace and complete the addition statement below.

Color in 7 hearts, then color in 5 more hearts to help find the sum.

7 + 5 =

WORKSHEET
ADDITION

1. Find the sum: 2 + 2 + 1

Trace and complete the addition statement below.

Color in 2 bowls, then 2 more bowls and another 1 bowl to help find the sum.

2 + 2 + 1 =

2. Find the sum: 3 + 2 + 3

Trace and complete the addition statement below.

Color in 3 triangles, then 2 more triangles and another 3 triangles to help find the sum.

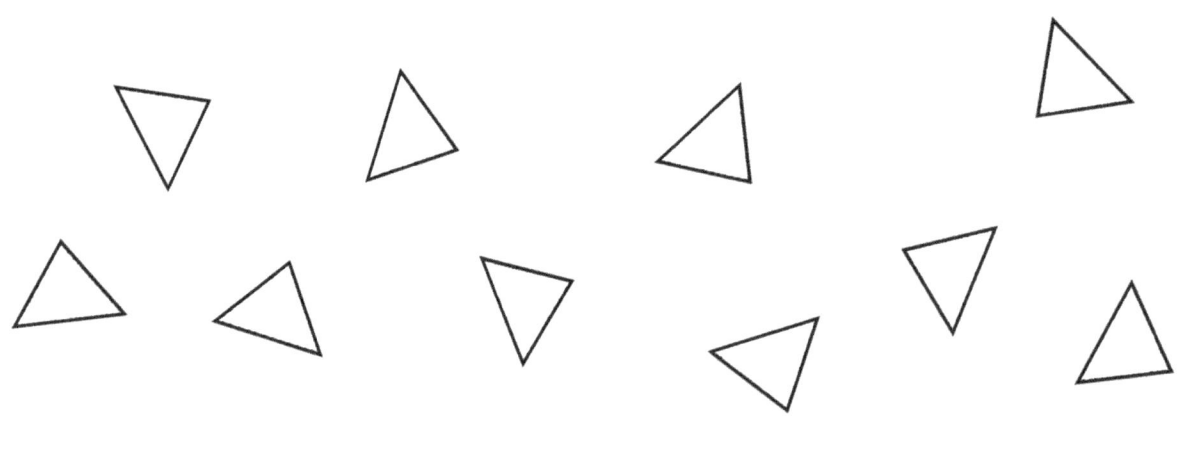

3 + 2 + 3 =

LESSON 11

The Commutative Property

THE COMMUTATIVE PROPERTY

Addition has the **commutative property**, which just means that the order in which you count the things you are adding does not matter.

Adding your 2 apples plus your friend's 1 apple: **2 + 1 = 3**

is the same as

adding your friend's 1 apple plus your 2 apples: **1 + 2 = 3**.

Either way, there are 3 apples all together!

Example: Find the sum 2 + 5, then switch the numbers around and find the sum 5 + 2

2 + 5: We color in **2** pears and **5** more, then count them.

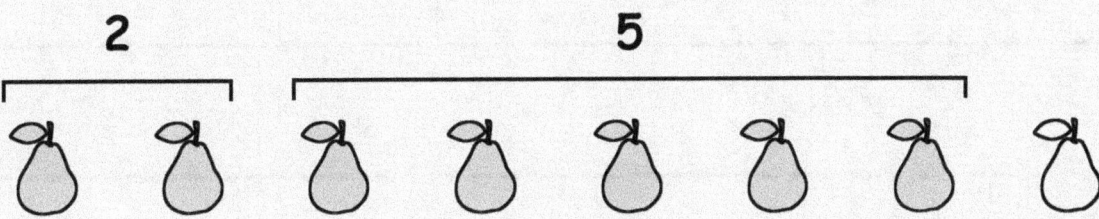

There are 7 pears colored in all together.

2 + 5 = 7

5 + 2: We color in **5** pears and **2** more, then count them.

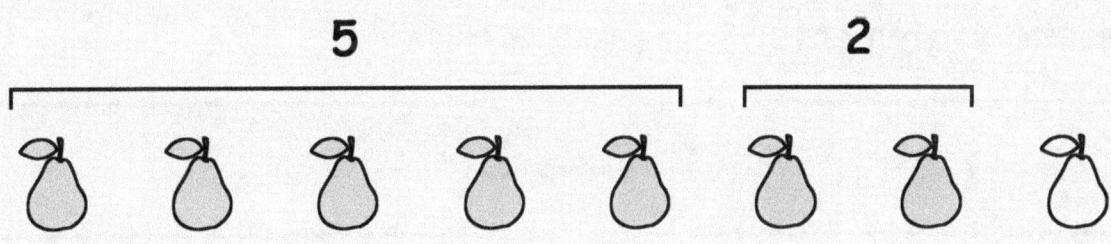

There are 7 pears colored in all together.

5 + 2 = 7

2 + 5 and 5 + 2 are both equal to 7.
The order of the numbers being added does not matter.

Trace the word, then write it on your own in the space below:

commutative

Example: Rewrite the addition statements.

Use the commutative property to rewrite the following addition statements by switching the positions of the numbers being added.

a) 2 + 4 = 6

Trace the statements:

2 + 4 = 6 → 4 + 2 = 6

b) 1 + 7 = 8

Trace the statements:

1 + 7 = 8 → 7 + 1 = 8

c) 3 + 4 = 7

Trace the statements:

d) 3 + 8 = 11

Trace the statements:

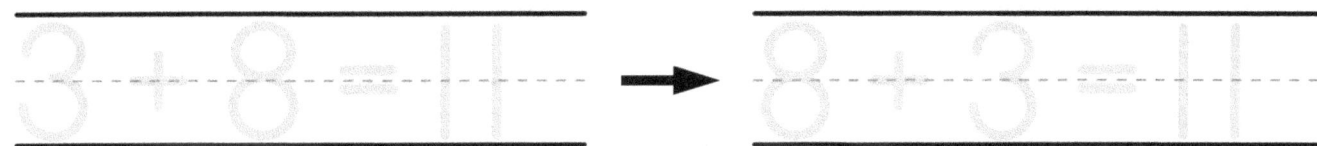

e) 5 + 10 = 15

Trace the statements:

WORKSHEET
THE COMMUTATIVE PROPERTY

Using the commutative property of addition, rewrite each addition statement by switching the positions of the numbers being added.

$2 + 3 = 5$ → $3 + 2 = 5$

$1 + 3 = 4$ →

$3 + 5 = 8$ →

$1 + 4 = 5$ →

$6 + 7 = 13$ →

$7 + 8 = 15$ →

WORKSHEET
THE COMMUTATIVE PROPERTY

Using the commutative property of addition, rewrite each addition statement by switching the positions of the numbers being added.

2 + 7 = 9 → 7 + 2 = 9

6 + 2 = 8 →

3 + 4 = 7 →

2 + 10 = 12 →

8 + 6 = 14 →

4 + 7 = 11 →

WORKSHEET
THE COMMUTATIVE PROPERTY

Using the commutative property of addition, rewrite each addition statement by switching the positions of the numbers being added.

$0 + 3 = 3$ → $3 + 0 = 3$

$1 + 7 = 8$ →

$3 + 6 = 9$ →

$6 + 4 = 10$ →

$2 + 8 = 10$ →

$4 + 8 = 12$ →

LESSON 12

Subtraction

SUBTRACTION

Subtraction is used to find how many things you have left after taking some away. The amount that is left is called the **difference**.

Suppose you have 3 candies and you give 1 candy to your friend. We use subtraction to find out how many candies are left.

One way to find the difference is to cross out the amount taken away, then count how many are left:

There are 2 left after you take away (subtract) 1 candy.

3 candies - 1 candy = 2 candies

The **minus sign** "-" is used to show we are subtracting candies. After the subtraction is the **equal sign** "=", then the **difference**.

Example: Find the difference: 5 - 2

This can be figured out by coloring the pictures below and by drawing an X through the amount being subtracted (or taken away) from those colored in.

First: We color in **5** apples, then draw an X through **2** of those that are colored in.

Next: Count the total number of **colored-in** apples that are left.

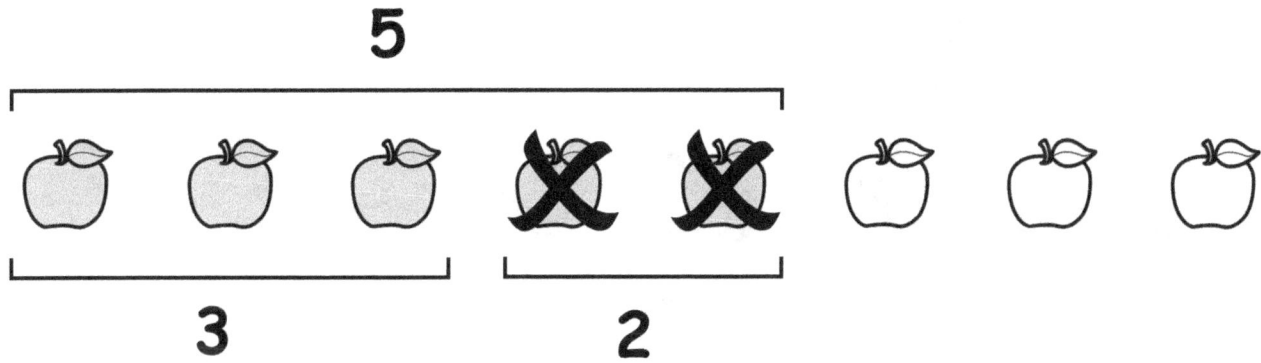

There are 3 apples left.

Trace the statement, then write it on your own in the space below:

5 - 2 = 3

Example: Find the difference: 4 - 2

This can be figured out by coloring the pictures below and by drawing an X through the amount being subtracted (or taken away) from those colored in.

First: We color in **4** ducks, then draw an X through **2** of those that are colored in.

Next: Count the total number of **colored-in** ducks that are left.

There are 2 ducks left.

Trace the statement, then write it on your own in the space below:

4 - 2 = 2

Example: Find the difference: 6 - 4

This can be figured out by coloring the pictures below and by drawing an X through the amount being subtracted (or taken away) from those colored in.

First: We color in **6** fish, then draw an X through **4** of those that are colored in.

Next: Count the total number of **colored-in** fish that are left.

There are 2 fish left.

Trace the statement, then write it on your own in the space below:

6 - 4 = 2

Trace the word, then write it on your own in the space below:

subtraction

Trace the word and minus signs, then write them on your own in the space below:

minus

Trace the word, then write it on your own in the space below:

difference

WORKSHEET
SUBTRACTION

1. Find the difference: 4 - 1

Trace and complete the subtraction statement below.

Color in 4 hearts, then draw an X through 1 colored-in heart to remove it. Count the remaining colored-in hearts to find the difference.

4 - 1 =

2. Find the difference: 6 - 3

Trace and complete the subtraction statement below.

Color in 6 balls, then draw an X through 3 colored-in balls to remove them. Count the remaining colored-in balls to find the difference.

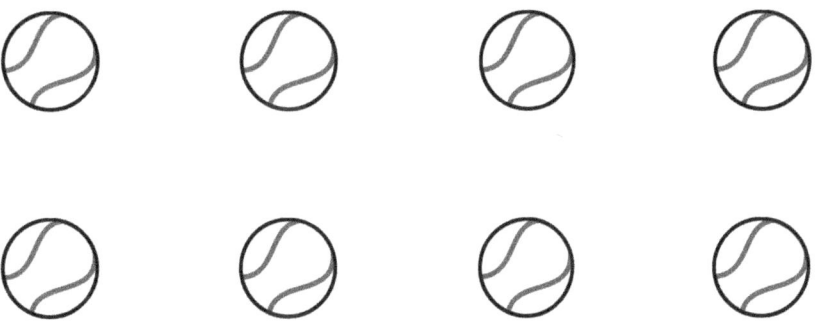

6 - 3 =

WORKSHEET
SUBTRACTION

1. Find the difference: 5 - 4

Trace and complete the subtraction statement below.

Color in 5 stars, then draw an X through 4 colored-in stars to remove them. Count the remaining colored-in stars to find the difference.

5 - 4 =

2. Find the difference: 7 - 2

Trace and complete the subtraction statement below.

Color in 7 smilies, then draw an X through 2 colored-in smilies to remove them. Count the remaining colored-in smilies to find the difference.

7 - 2 =

WORKSHEET
SUBTRACTION

1. Find the difference: 10 - 4

Trace and complete the subtraction statement below.

Color in 10 beach balls, then draw an X through 4 colored-in beach balls to remove them. Count the remaining colored-in balls to find the difference.

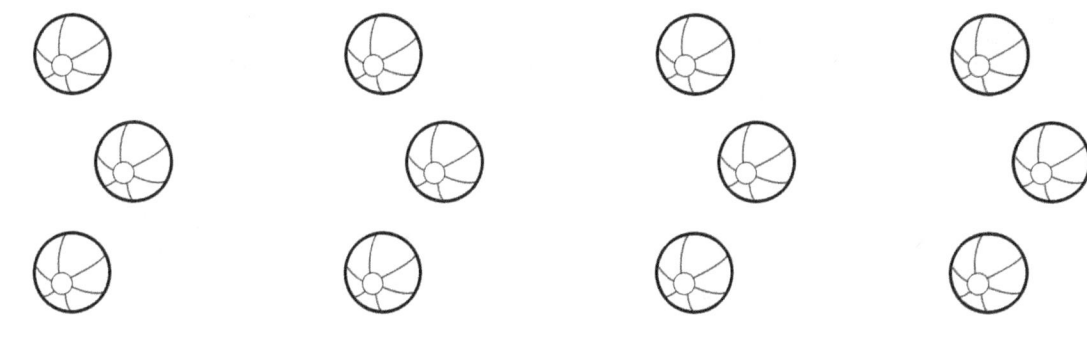

10 - 4 =

2. Find the difference: 12 - 5

Trace and complete the subtraction statement below.

Color in 12 jelly beans, then draw an X through 5 colored-in jelly beans to remove them. Count the remaining colored-in jelly beans to find the difference.

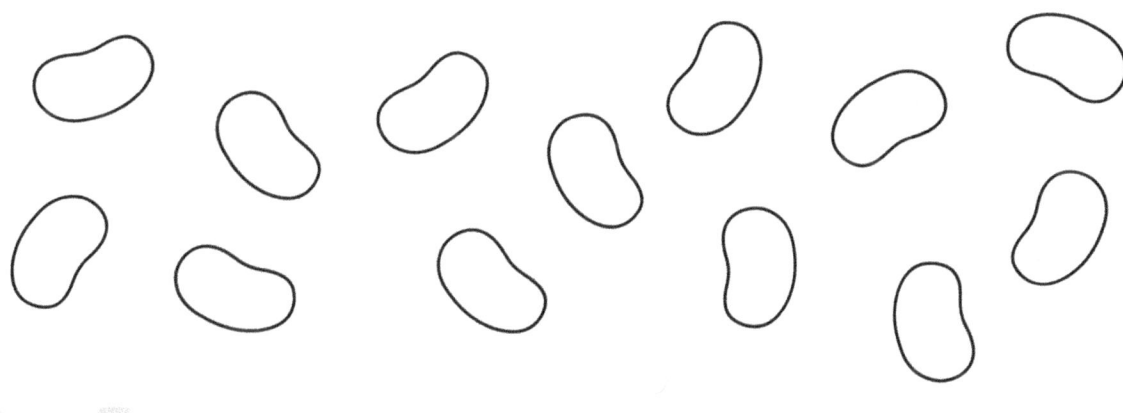

12 - 5 =

WORKSHEET
SUBTRACTION

1. Find the difference: 5 - 0

Trace and complete the subtraction statement below.

Color in 5 pinwheeels, then draw an X through 0 colored-in pinwheels. Count the remaining colored-in pinwheels to find the difference.

5 - 0 =

2. Find the difference: 4 - 4

Trace and complete the subtraction statement below.

Color in 4 turtles, then draw an X through 4 colored-in turtles to remove them. Count the remaining colored-in turtles to find the difference.

4 - 4 =

LESSON 13

Tables: Rows & Columns

TABLES: ROWS & COLUMNS

Tables are used to organize information in a way that is easy to read.

Tables are made of rows and columns.
    Columns go down the page.
    Rows go across the page.

Below is an example of a table.
Each column has a **label**: Lunch and Toys.
Each row has a **label**: Abby, Chris and Mark.

	Lunch	Toys
Abby	pizza	duck
Chris	bowl	teddy bear
Mark	orange and pear	pinwheel

Trace the words, then write them on your own in the spaces below:

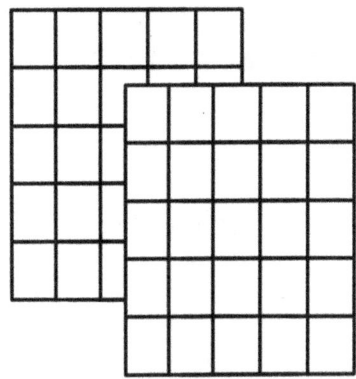
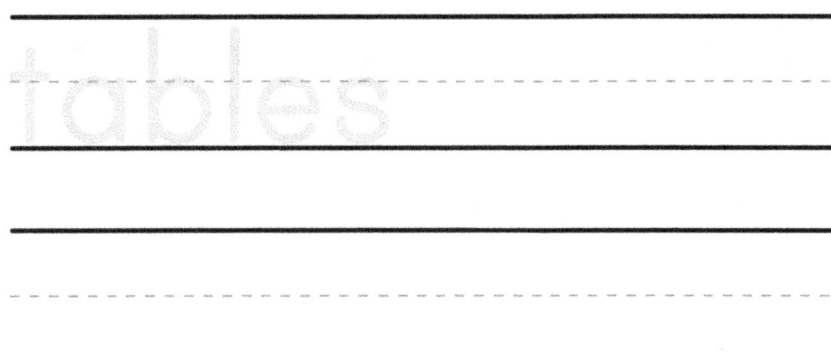

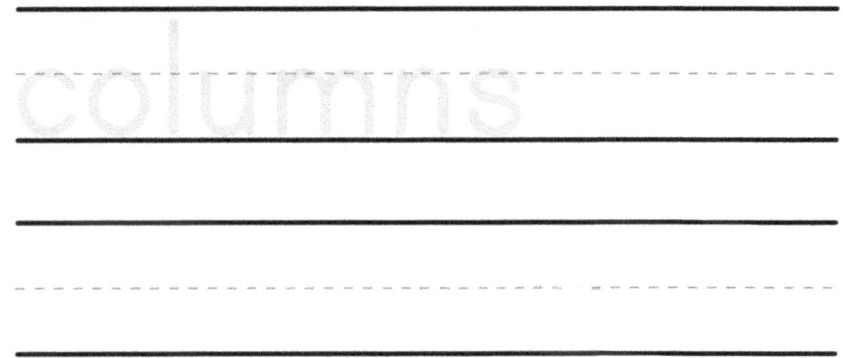

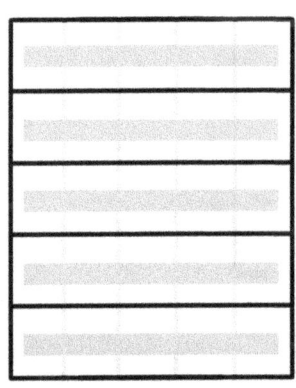

Example: Use the table below to find out which toy belongs to Mark.

First, we find the **column** named **Toys** and draw a line DOWN the **column**.

Next, we find the **row** named **Mark** and draw a line ACROSS the **row**.

The box where the lines cross is the one with Mark's toy.

Trace the grey lines drawn on the table.

	Lunch	Toys
Abby	pizza	duck
Chris	bowl	teddy bear
Mark	apple & pear	pinwheel

Mark's toy is a pinwheel.

Example: Use the table below to find out who has fruit for lunch.

First, we find the **Lunch column** and draw a line DOWN the column until we reach the fruit.

Next, we draw a line from the fruit, ACROSS to the left until we reach the correct name.

Trace the grey lines drawn on the table.

	Lunch	Toys
Abby	pizza	duck
Chris	soup	teddy bear
Mark	fruit	pinwheel

Mark has fruit for lunch.

WORKSHEET
TABLES: ROWS & COLUMNS

Use the table to answer the questions below.

	Lunch	Toys
Abby	🍕	🦆
Chris	🥣	🧸
Mark	🍊🍐	🎐

Circle the correct answers.

1. Which lunch belongs to Abby?   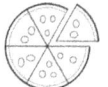      

2. Who has 🥣 for lunch?   Abby   Chris   Mark

3. Which toy belongs to Abby?

4. Whose toy is a 🎐 ?   Abby   Chris   Mark

WORKSHEET
TABLES: ROWS & COLUMNS

Kevin, Tina and Mary each have a flower garden. The table below shows the amounts and types of flowers they are growing.

	🌷	🌼	🌸	🌱
Kevin	2	1	3	2
Tina	4	0	2	1
Mary	3	3	1	0

1. How many 🌼 does Mary have? _____

2. How many 🌱 does Kevin have? _____

3. How many 🌸 does Tina have? _____

4. How many 🌼 does Tina have? _____

WORKSHEET
TABLES: ROWS & COLUMNS

Kevin, Tina and Mary each have a flower garden. The table below shows the amounts and types of flowers they are growing.

	🌷	🌼	🌸	🌷
Kevin	2	1	3	2
Tina	4	0	2	1
Mary	3	3	1	0

1. Who has 2 🌸 ? _____

2. Who has 3 🌼 ? _____

3. Who has no 🌷 ? _____

4. Who has 2 🌷 ? _____

WORKSHEET
TABLES: ROWS & COLUMNS

Kevin, Tina and Mary each have a flower garden. The table below shows the amounts and types of flowers they are growing.

	🌷	🌼	🌸	🌷
Kevin	2	1	3	2
Tina	4	0	2	1
Mary	3	3	1	0

1. How many 🌷 do Kevin and Tina have all together? _____

2. How many flowers does Mary have in her garden? _____

3. How many 🌷 do Kevin, Tina and Mary have all together? _____

WORKSHEET
TABLES: ROWS & COLUMNS

Kevin, Tina and Mary each have a flower garden. The table below shows the amounts and types of flowers they are growing.

	🌷	🌻	🌼	🌷
Kevin	2	1	3	2
Tina	4	0	2	1
Mary	3	3	1	0

1. How many 🌼 do Mary and Tina have all together? _____

2. How many flowers does Kevin have in his garden? _____

3. How many 🌷 do Kevin, Tina and Mary have all together? _____

LESSON 14

Counting (Up to 100)

COUNTING (UP TO 100)

So far we have counted up to 20. Now let's count past 20 and up to 100!

It's Easy!

To get the numbers 10 through 19, you wrote a 1 in front of each number 0 through 9:

10   11   12   13   14   15   16   17   18   19

Counting the other numbers up to 100 works the same way.

Let's start with the 20's:

20 through 29

We write a 2 in front of each of the numbers 0 through 9:

20   21   22   23   24   25   26   27   28   29

Trace the numbers 20 through 29:

20   21   22   23   24
25   26   27   28   29

30 through 39

We write a 3 in front of each of the numbers 0 through 9:

30 31 32 33 34 35 36 37 38 39

Trace the numbers 30 through 39:

30 31 32 33 34
35 36 37 38 39

40 through 49

We write a 4 in front of each of the numbers 0 through 9:

40 41 42 43 44 45 46 47 48 49

Trace the numbers 40 through 49:

40 41 42 43 44
45 46 47 48 49

50 through 59

We write a 5 in front of each of the numbers 0 through 9:

50 51 52 53 54 55 56 57 58 59

Trace the numbers 50 through 59:

50 51 52 53 54

55 56 57 58 59

60 through 69

We write a 6 in front of each of the numbers 0 through 9:

60 61 62 63 64 65 66 67 68 69

Trace the numbers 60 through 69:

60 61 62 63 64

65 66 67 68 69

70 through 79

We write a 7 in front of each of the numbers 0 through 9:

70　71　72　73　74　75　76　77　78　79

Trace the numbers 70 through 79:

70　71　72　73　74
75　76　77　78　79

80 through 89

We write an 8 in front of each of the numbers 0 through 9:

80　81　82　83　84　85　86　87　88　89

Trace the numbers 80 through 89:

80　81　82　83　84
85　86　87　88　89

90 through 99

We write a 9 in front of each of the numbers 0 through 9:

90 91 92 93 94 95 96 97 98 99

Trace the numbers 90 through 99:

90 91 92 93 94

95 96 97 98 99

And finally 100

Trace the number and number name, then write it on your own in the space below:

100   one hundred

Trace the numbers and names then write them on your own in the blanks below:

20 twenty

30 thirty

40 forty

50 fifty

Trace the numbers and names then write them on your own in the blanks below:

60   sixty

70   seventy

80   eighty

90   ninety

Here are the numbers 0-100 all together.
These are in order from smallest to largest.

```
0    1    2    3    4    5    6    7    8    9
10   11   12   13   14   15   16   17   18   19
20   21   22   23   24   25   26   27   28   29
30   31   32   33   34   35   36   37   38   39
40   41   42   43   44   45   46   47   48   49
50   51   52   53   54   55   56   57   58   59
60   61   62   63   64   65   66   67   68   69
70   71   72   73   74   75   76   77   78   79
80   81   82   83   84   85   86   87   88   89
90   91   92   93   94   95   96   97   98   99
100
```

WORKSHEET
COUNTING (UP TO 100)

Trace each group of numbers and fill in the blank with the number that comes AFTER them.

10 11 ___   27 28 ___

22 23 ___   31 32 ___

45 46 ___   84 85 ___

67 68 ___   77 78 ___

59 60 ___   49 50 ___

78 79 ___   28 29 ___

WORKSHEET
COUNTING (UP TO 100)

Trace each group of numbers and fill in the blank with the number that comes BEFORE them.

____ 12 13   ____ 17 18

____ 24 25   ____ 41 42

____ 32 33   ____ 82 83

____ 44 45   ____ 67 68

____ 59 60   ____ 29 30

____ 80 81   ____ 70 71

WORKSHEET
COUNTING (UP TO 100)

Trace each group of numbers and fill in the blank with the number that comes BETWEEN them.

14 ___ 16   20 ___ 22

42 ___ 44   35 ___ 37

27 ___ 29   62 ___ 64

87 ___ 89   51 ___ 53

78 ___ 80   18 ___ 20

59 ___ 61   49 ___ 51

WORKSHEET
COUNTING (UP TO 100)

Trace the numbers in each row and fill in the blanks with the missing numbers.

20 ___ 22 ___ 24

___ 36 ___ 38 39

55 ___ ___ 58 59

44 45 ___ 47 ___

___ ___ 19 20 21

77 78 ___ ___ 81

___ 30 31 ___ 33

LESSON 15

Counting by 2's

COUNTING BY 2's

Counting by 1's

When you count each number without skipping any, you are counting by 1's.

1 2 3 4 5 6 7 8 9 10 11 12 13 14 15 16 17 18 19 20

Sometimes it's easier to count **groups** of things instead of 1 at a time.

Counting by 2's

To count by 2's, start at 2 and skip a number each time you count.

The skipped numbers are crossed out:

~~1~~ 2 ~~3~~ 4 ~~5~~ 6 ~~7~~ 8 ~~9~~ 10

~~11~~ 12 ~~13~~ 14 ~~15~~ 16 ~~17~~ 18 ~~19~~ 20

Example: Count the frogs by 1's and then by 2's.

Let's count the frogs below by 1's (1 at a time):

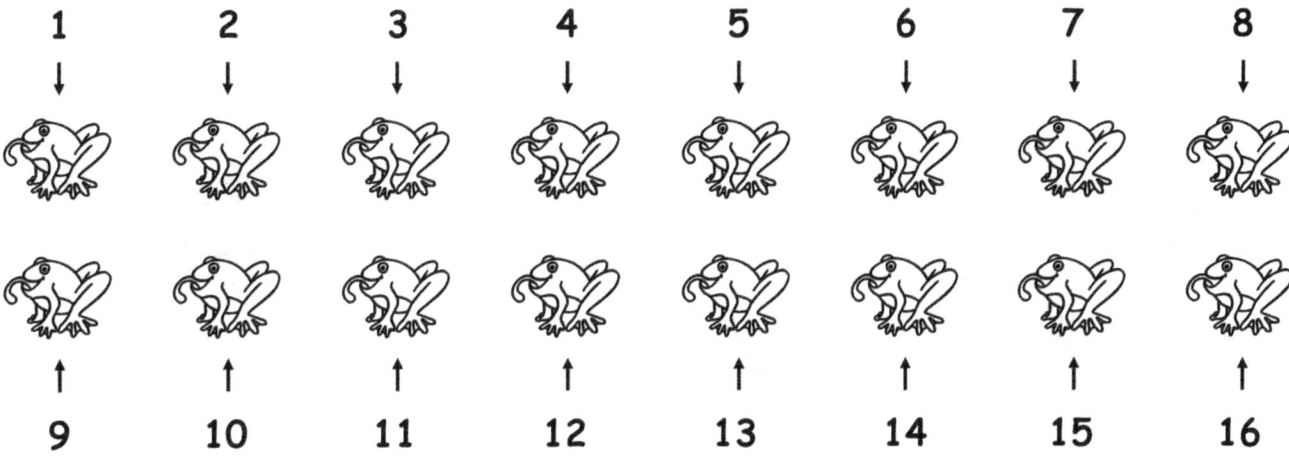

Now, let's count the frogs below by 2's to count 2 at a time.
Trace the numbers in the boxes.

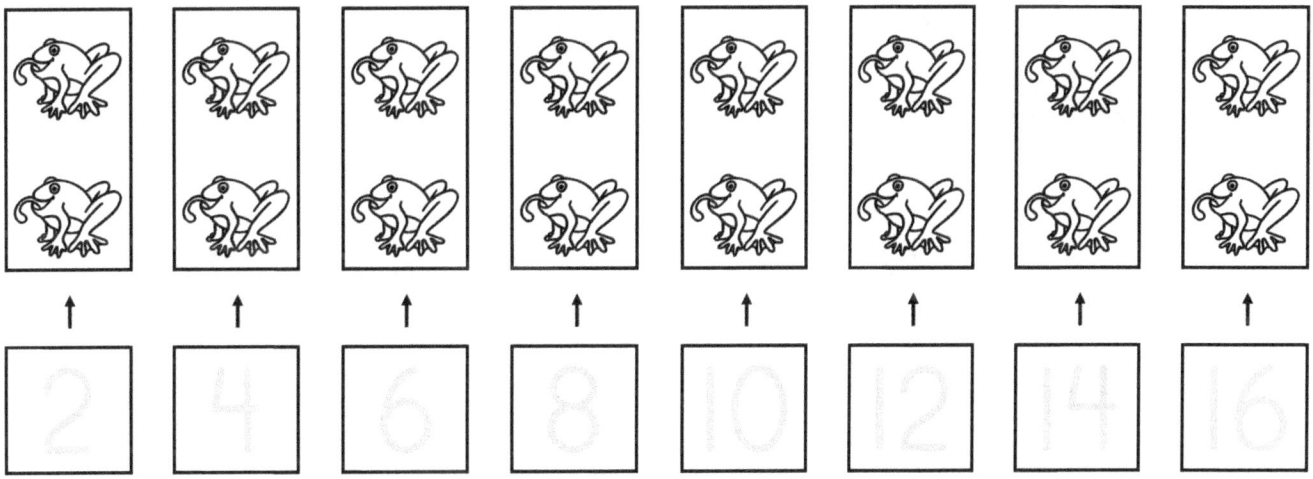

It's a lot faster to count all of these frogs by 2's!

Example: Count the ladybugs by 2's:

The ladybugs below were circled 2 at a time. Going in order, from left to right, trace the numbers next to each pair of ladybugs to count them by 2's.

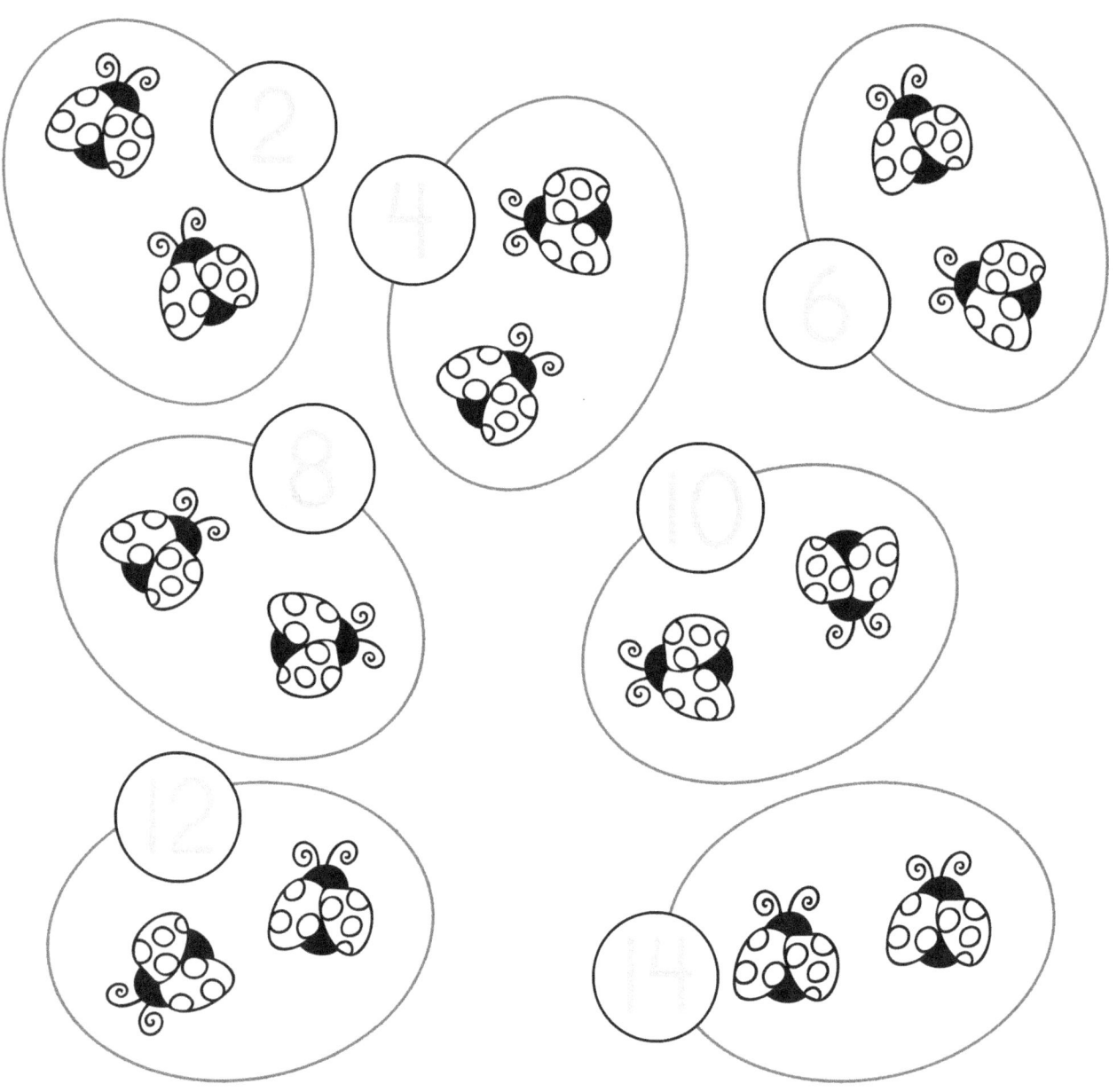

There are 14 ladybugs.

WORKSHEET
COUNTING BY 2's

1. Count the monkeys by 2's and write the total amount in the blank below.

How many monkeys? _____

2. Count the chicks by 2's and write the total amount in the blank below.

How many chicks? _____

3. Count the balls by 2's and write the total amount in the blank below.

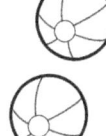 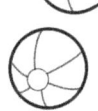 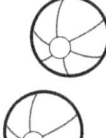  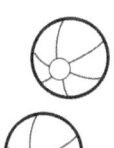  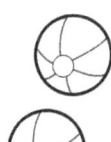 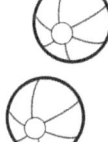

How many balls? _____

WORKSHEET
COUNTING BY 2's

1. Count the oranges below by 2's and write the total amount in the blank below.

How many oranges? _____

2. Count the pinwheels below by 2's and write the total amount in the blank below.

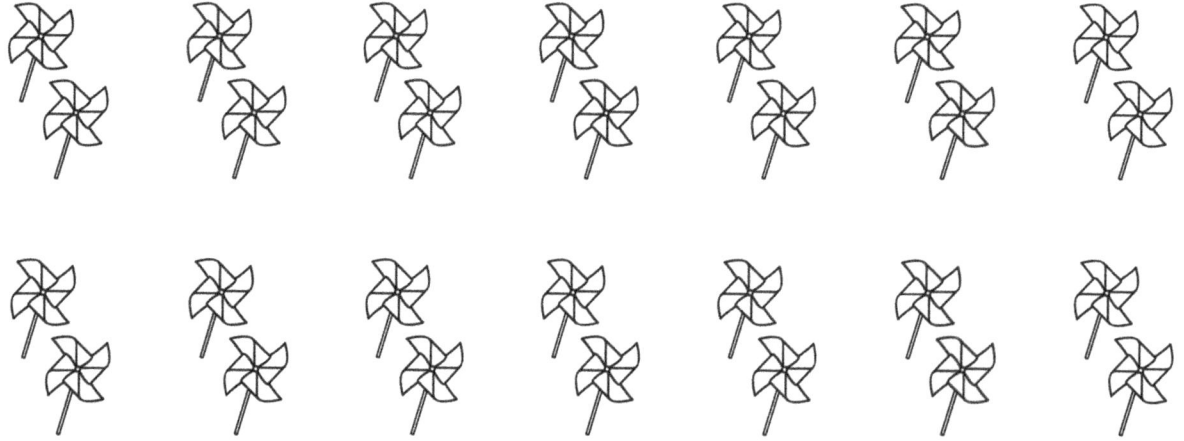

How many pinwheels? _____

WORKSHEET
COUNTING BY 2's

1. Count the mice below by 2's and write the total amount in the blank below.

How many mice? _____

2. Count the flowers below by 2's and write the total amount in the blank below.

How many flowers? _____

LESSON 16

Counting by 10's

COUNTING BY 10's

To count by 10's, start at 10 and count every 10th number.

When counting in this way, these are just the numbers that end in 0.

| 10 | 20 | 30 | 40 | 50 |
| 60 | 70 | 80 | 90 | 100 |

Trace the numbers 10 through 100 counting by 10's, then write them on your own in the spaces below:

10 20 30 40 50

_____

60 70 80 90 100

_____

Here are the numbers 1 through 100 with the numbers counting by 10's circled.

They are all the numbers that end in 0.

1	2	3	4	5	6	7	8	9	**10**
11	12	13	14	15	16	17	18	19	**20**
21	22	23	24	25	26	27	28	29	**30**
31	32	33	34	35	36	37	38	39	**40**
41	42	43	44	45	46	47	48	49	**50**
51	52	53	54	55	56	57	58	59	**60**
61	62	63	64	65	66	67	68	69	**70**
71	72	73	74	75	76	77	78	79	**80**
81	82	83	84	85	86	87	88	89	**90**
91	92	93	94	95	96	97	98	99	**100**

Example: Count the rabbits by 10's.

There are 10 rabbits in each row. Trace the numbers in the boxes next to each row to count them by 10's.

Example: Count the dragonflies by 10's.

There are 10 dragonflies in each row. Trace the numbers in the boxes next to each row to count them by 10's.

WORKSHEET
COUNTING BY 10's

1. There are 10 flowers in each box. Count them by 10's and write the total amount in the blank below.

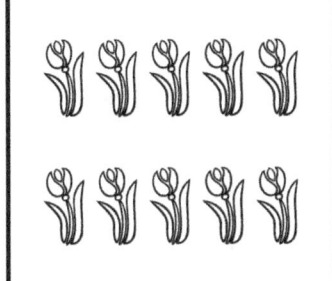 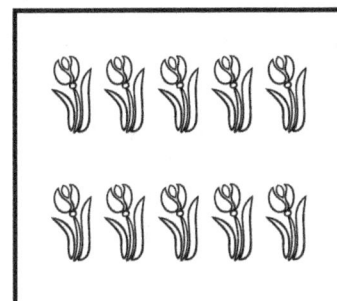 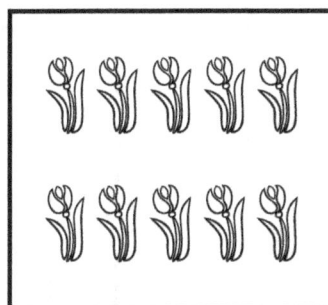

How many flowers? _____

2. There are 10 candies in each row. Count them by 10's and write the total amount in the blank below.

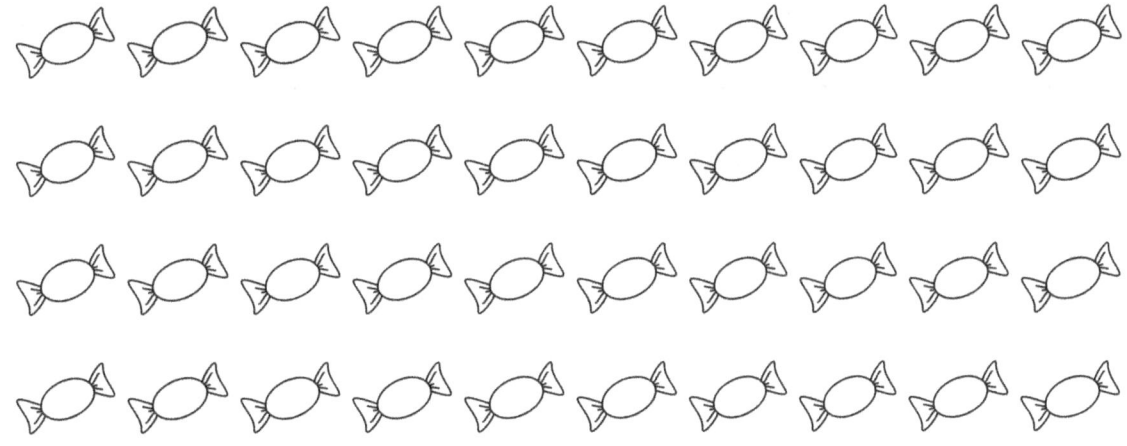

How many candies? _____

WORKSHEET
COUNTING BY 10's

1. There are 10 pears in each box. Count them by 10's and write the total amount in the blank below.

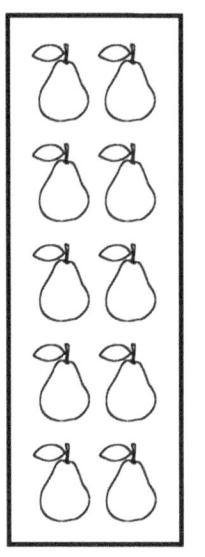 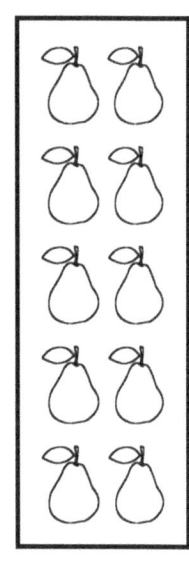 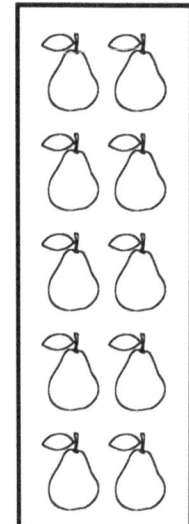 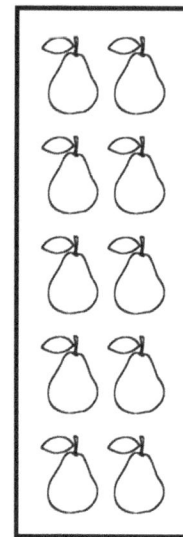 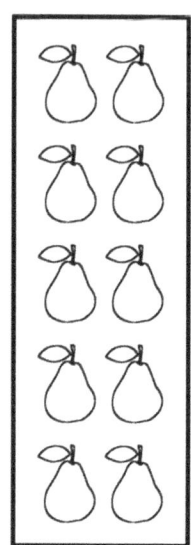

How many pears? _____

2. There are 10 flowers in each box. Count them by 10's and write the total amount in the blank below.

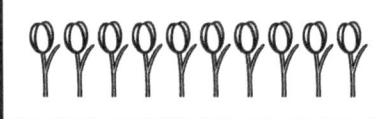 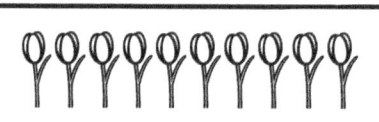 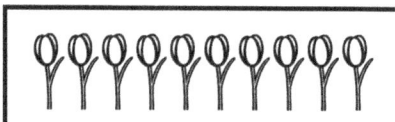

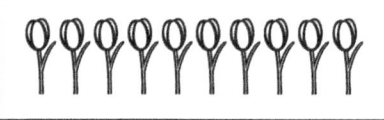 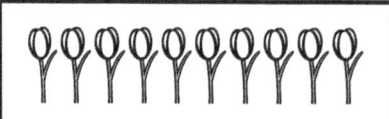 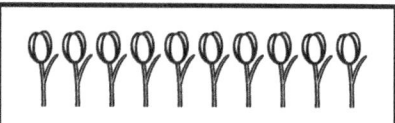

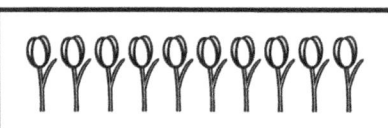

How many flowers? _____

WORKSHEET
COUNTING BY 10's

There are 10 balls in each column. Count them by 10's and write the total amount in the blank below.

How many balls? _____

LESSON 17

Counting by 5's

COUNTING BY 5's

To count by 5's, start at 5 and count every 5th number.

These are the numbers that end in either 5 or 0.

| 5 | 10 | 15 | 20 | 25 |

| 30 | 35 | 40 | 45 | 50 |

Trace the numbers 5 through 50 counting by 5's, then write them on your own in the spaces below:

5   10   15   20   25

30   35   40   45   50

Here are the numbers 1 through 100 with the numbers counting by 5's circled.

This one is similar to the chart for counting by 10's, but also includes the column of numbers that end in 5.

1	2	3	4	5	6	7	8	9	10
11	12	13	14	15	16	17	18	19	20
21	22	23	24	25	26	27	28	29	30
31	32	33	34	35	36	37	38	39	40
41	42	43	44	45	46	47	48	49	50
51	52	53	54	55	56	57	58	59	60
61	62	63	64	65	66	67	68	69	70
71	72	73	74	75	76	77	78	79	80
81	82	83	84	85	86	87	88	89	90
91	92	93	94	95	96	97	98	99	100

Example: Count the caterpillars by 5's.

There are 5 caterpillars in each column. Trace the numbers in the boxes below each column to count them by 5's.

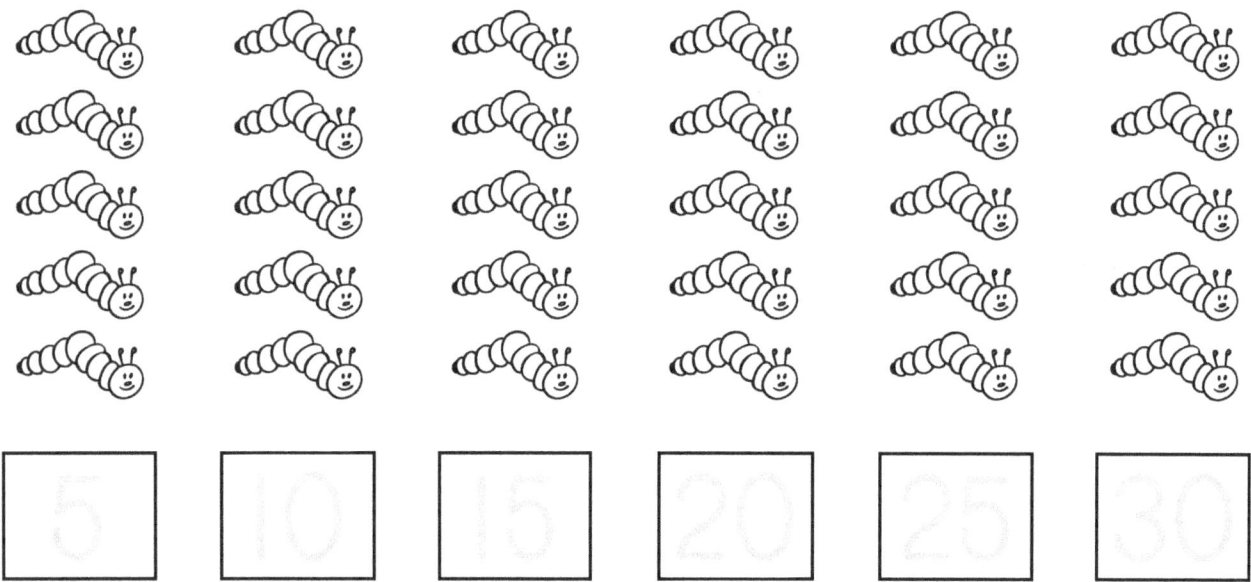

Example: Count the flowers by 5's.

There are 5 flowers in each set. Trace the numbers in the boxes next to each row to count them by 5's.

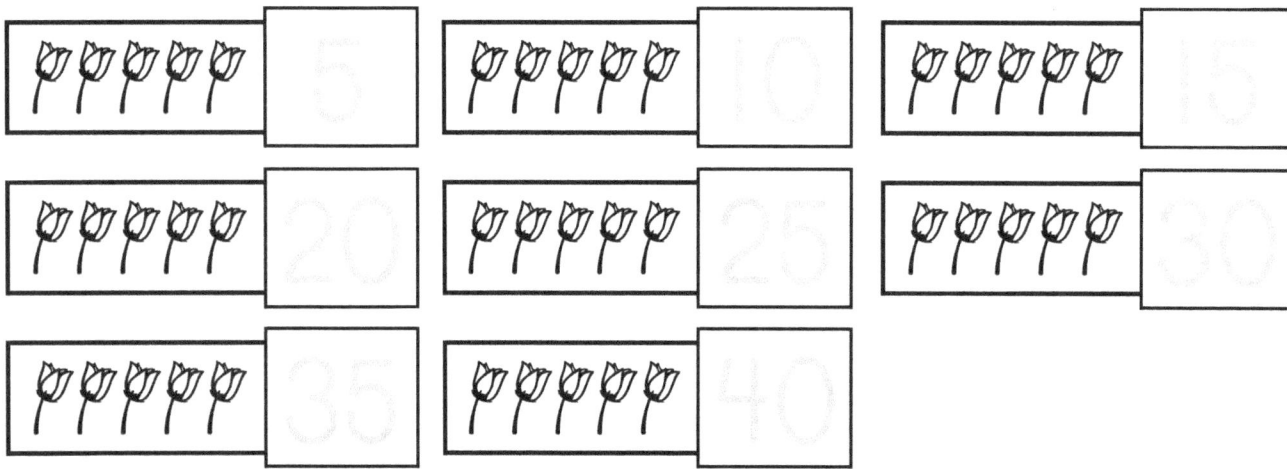

WORKSHEET
COUNTING BY 5's

1. There are 5 turtles in each column. Count them by 5's and write the total amount in the blank below.

How many turtles? _____

2. There are 5 apples in each box. Count them by 5's and write the total amount in the blank below.

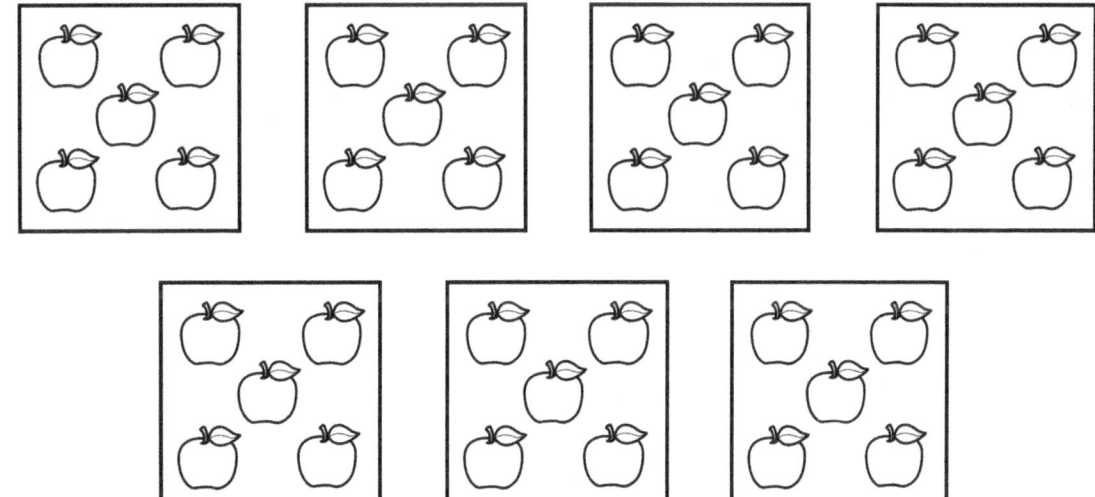

How many apples? _____

WORKSHEET
COUNTING BY 5's

1. There are 5 fish in each fishbowl. Count them by 5's and write the total amount in the blank below.

How many fish? _____

2. There are 5 flowers in each group. Count them by 5's and write the total amount in the blank below.

How many flowers? _____

WORKSHEET
COUNTING BY 5's

There are 5 ladybugs in each circle. Count them by 5's and write the total amount in the blank below.

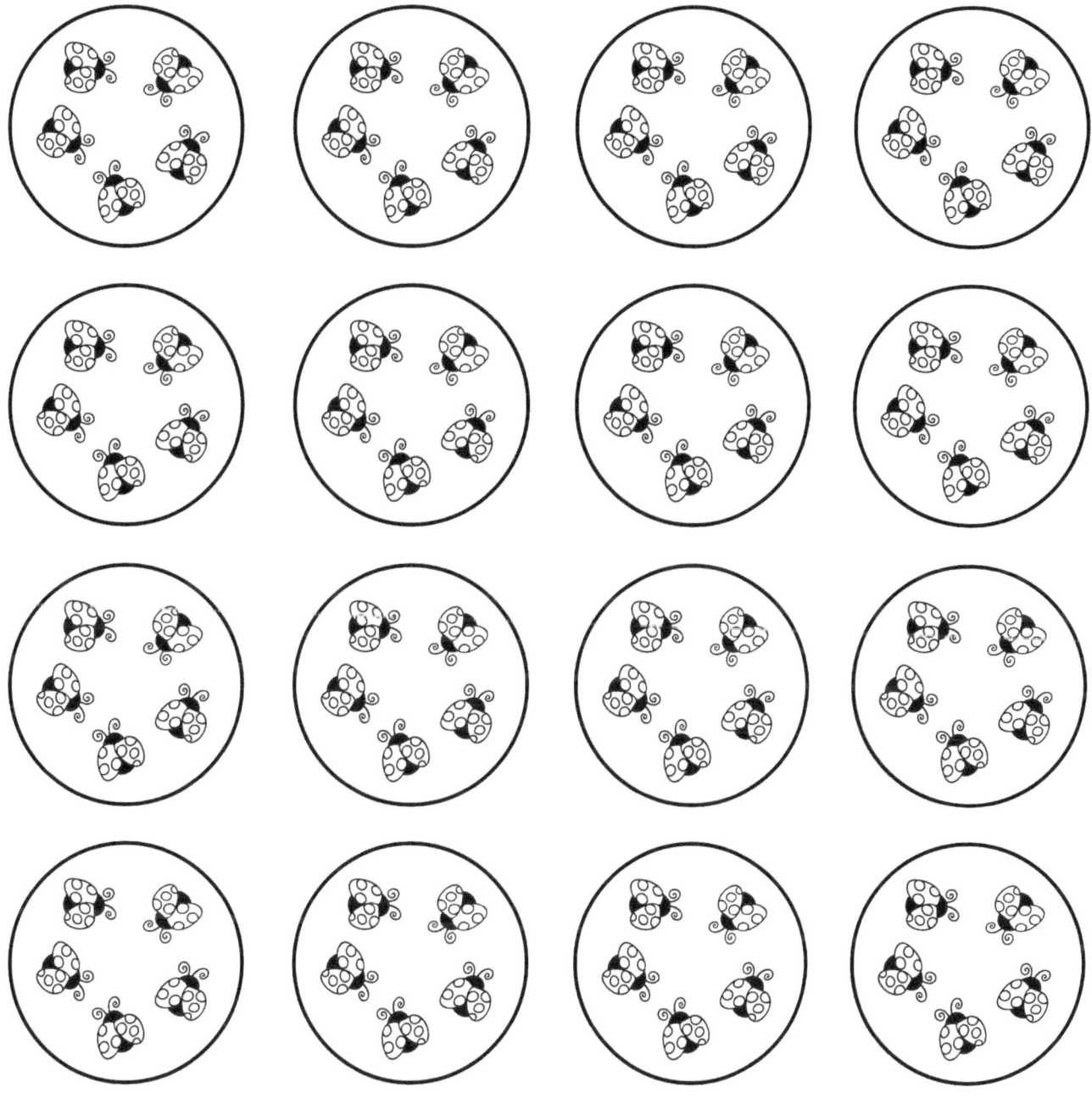

How many ladybugs? _____

LESSON 18

Tallying

TALLYING

Tallying is used when counting several different things at once using **tally marks** to keep track of the count.

A **tally mark** is a small line that is drawn each time something is counted.

Every 5th tally mark is drawn across the 4 before it. Then a new group is started after that.

Below is a tally count of 19 people's favorite colors.
This tallying helps to keep track of the 4 favorite colors in 1 chart.

Trace the tally marks.

Favorite Color	Tally Count	
RED	|||	3 people like RED
BLUE	⤫⤫ ||	7 people like BLUE
GREEN	⤫⤫	5 people like GREEN
PURPLE	||||	4 people like PURPLE

Example: Tally the shapes

Let's tally the shapes below.

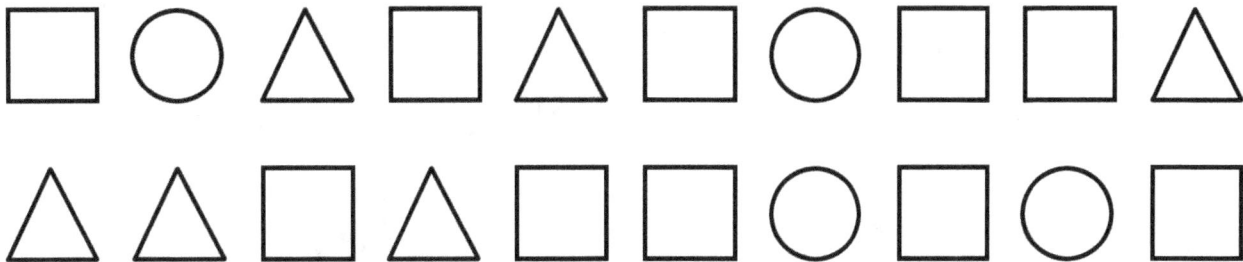

In each row, going from left to right, cross out 1 shape at a time. Each time you cross one out, trace a tally mark next to that shape in the chart below.

Do this for each of the shapes until they are all crossed out and all of the tally marks are traced.

Trace the tally marks below.

◯ Circle	||||	4 circles
▢ Square	卌 卌	10 squares
△ Triangle	卌 |	6 triangles

Example: Count the tally marks.

Tally marks are grouped into sets of 5 marks or less. This makes them easy to count - we just count by 5's and then count the single ones by 1's.

Trace the numbers below to count the tally marks

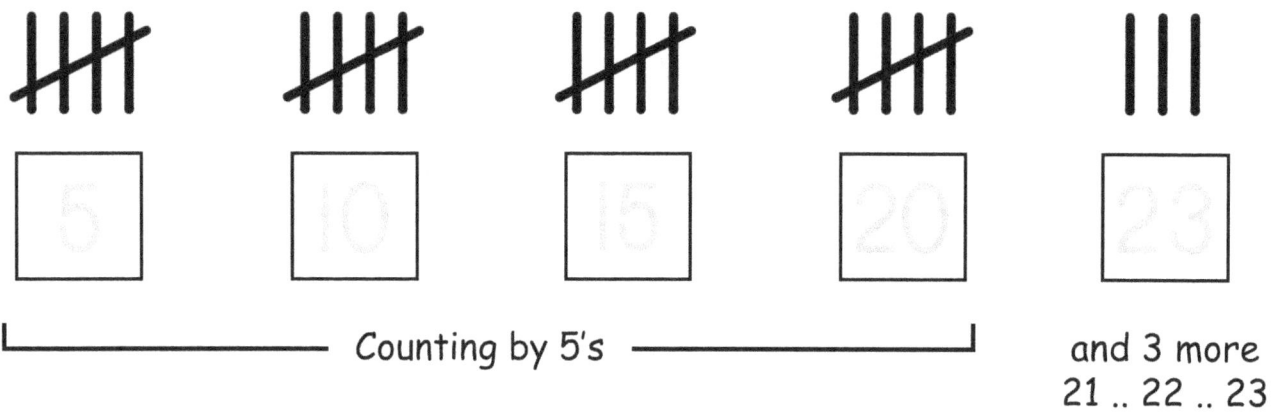

There are 23 tally marks all together.

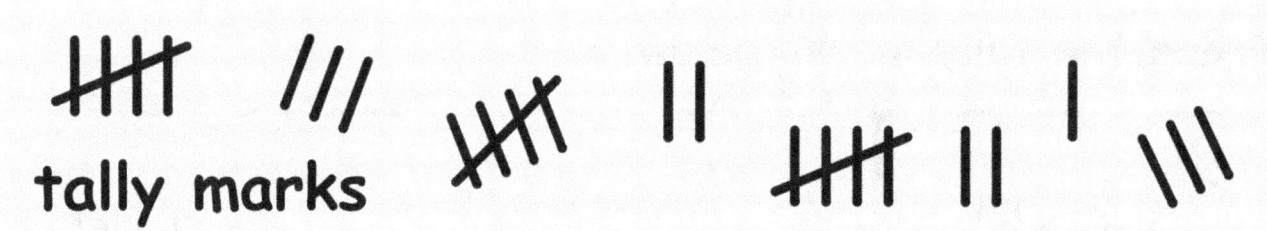

Trace the words, then write them on your own in the space below:

WORKSHEET
TALLYING

Tally the different critter types below.
Write the tally marks in the **tally marks** column. Then count the tally marks for each critter type and write the number amount in the **amount** column.

critters	tally marks	amount
🐸		
🐛		
🐞		

WORKSHEET
TALLYING

Tally the different animal types below.
Write the tally marks in the **tally marks** column. Then count the tally marks for each animal type and write the number amount in the **amount** column.

animals	tally marks	amount
🐵		
🦉		
🐤		

WORKSHEET
TALLYING

Tally the different flower types below.
Write the tally marks in the **tally marks** column. Then count the tally marks for each flower type and write the number amount in the **amount** column.

flowers	tally marks	amount
🌷		
✳		
🌸		

WORKSHEET
TALLYING

Count the tally marks.
Below is a tally chart of favorite ice cream toppings. For each topping, count the tally marks and write the number amount in the amount column.

toppings	tally marks	amount
hot fudge	|||| |||| |||| |||| ||||	
caramel	|||| |||| |||	
whipped cream	|||| |||| ||||		

Which topping was favored most?
_____
_____

Which topping was favored least?
_____
_____

WORKSHEET
TALLYING

Count the tally marks.

Below is a tally chart of favorite fruits. For each fruit, count the tally marks and write the number amount in the amount column.

fruit	tally marks	amount
orange	|||| |||| |||| ||||	
banana	|||| |||| ||||	
watermelon	|||| |||| |||| |||| |||	

Which fruit was favored most?
_____
_____

Which fruit was favored least?
_____
_____

WORKSHEET
TALLYING

Count the tally marks.
Below is a tally chart of favorite sports. For each sport, count the tally marks and write the number amount in the amount column.

sports	tally marks	amount
basketball	|||| |||| |||| ||||	
baseball	|||| |||| |||| ||||	
football	|||| |||| |||| |||| ||	

Which sport was favored most?
_____

Which sport was favored least?
_____

SOLUTIONS

Mastering the Math Milestone K⁺

Page 9

Page 10

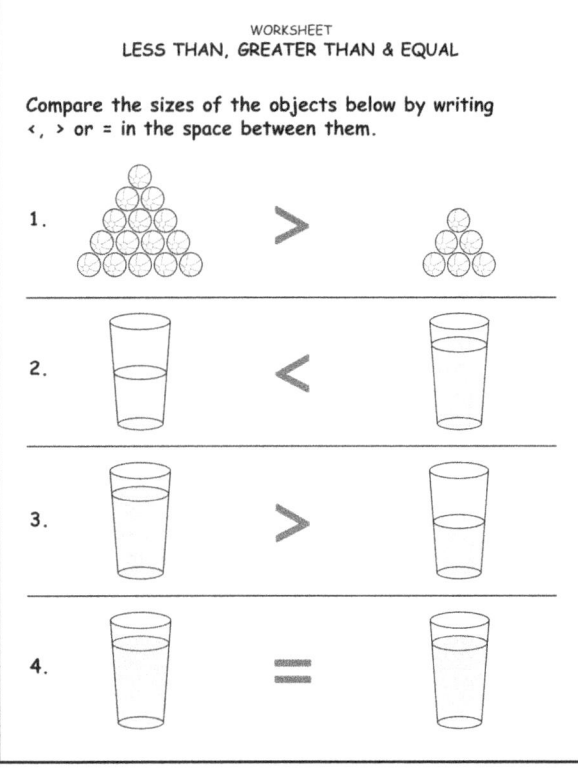

Page 11

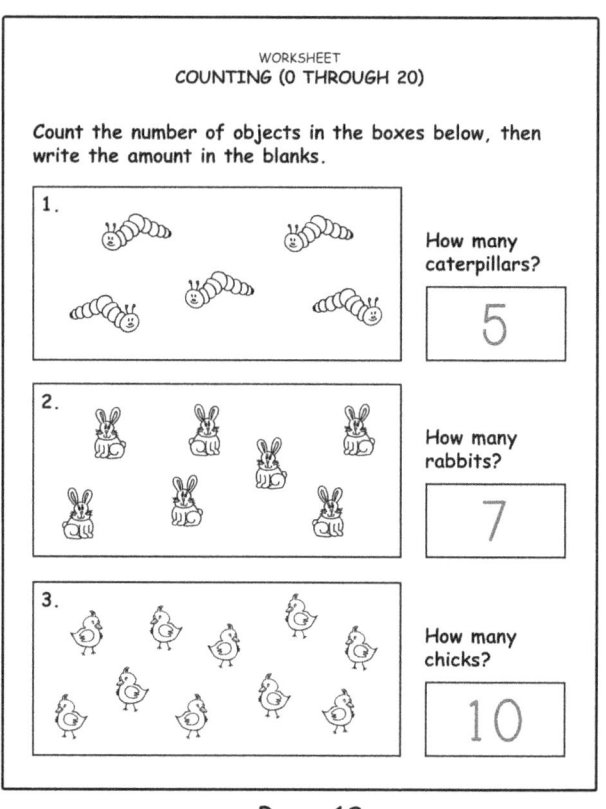

Page 19

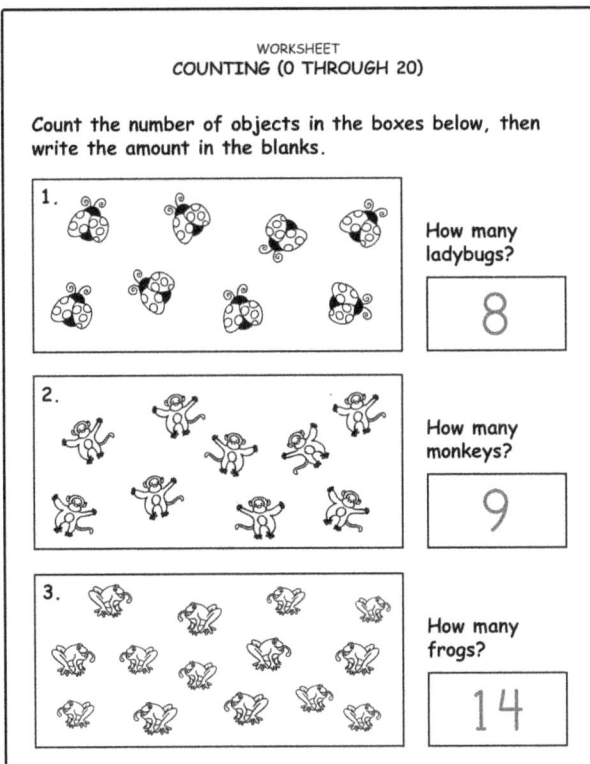

Page 20

WORKSHEET
COUNTING (0 THROUGH 20)

Count the number of objects in the boxes below, then write the amount in the blanks.

1. How many mice? 12
2. How many flower pots? 16
3. How many flowers? 18

Page 21

Page 22

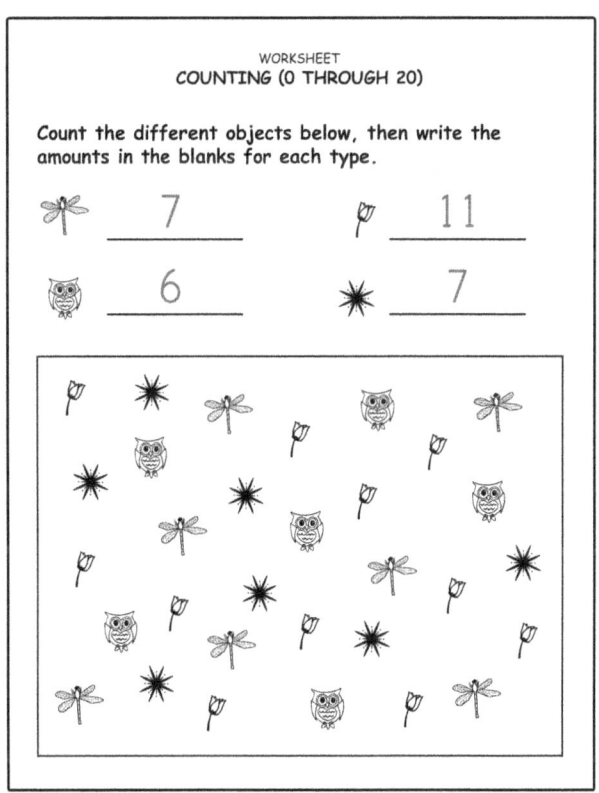

Page 23

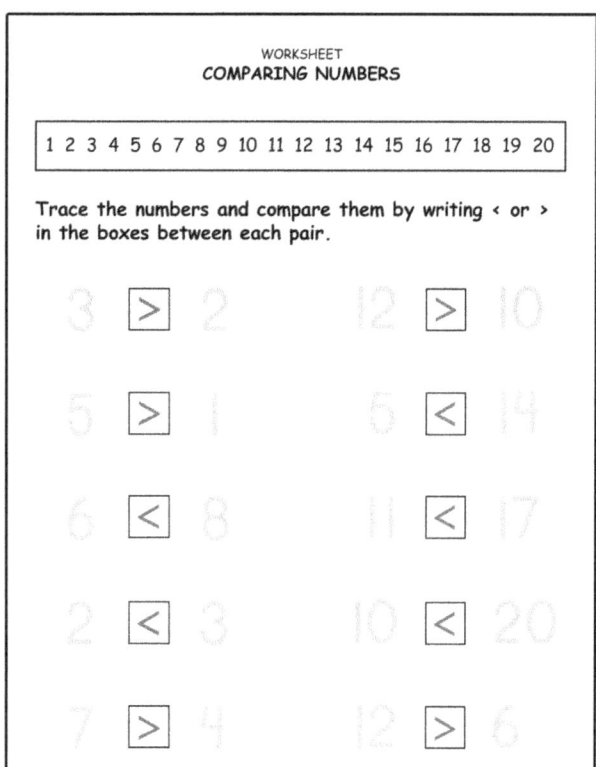

Page 29

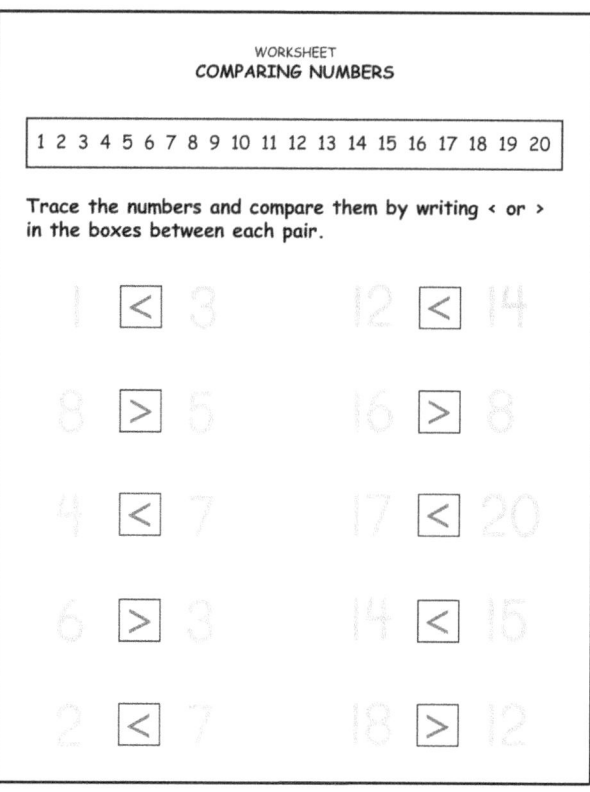

Page 30

Page 31

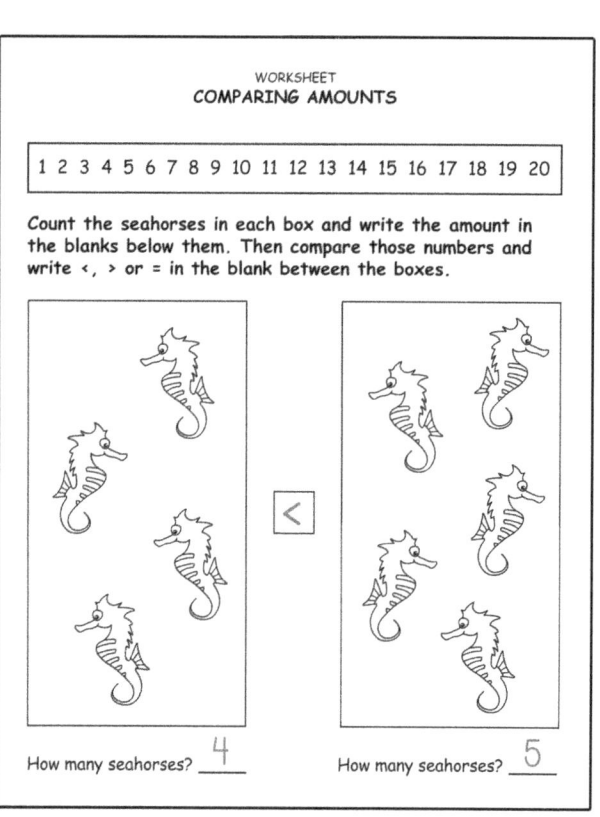

Page 36

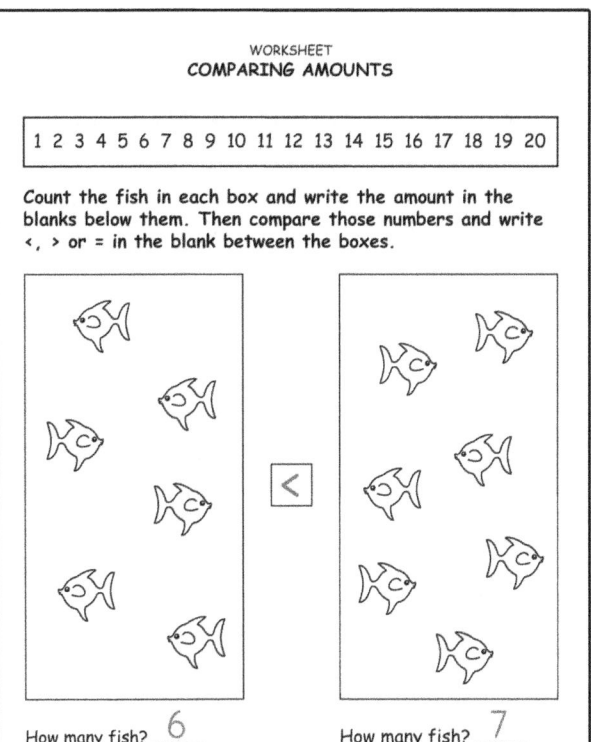

Page 37

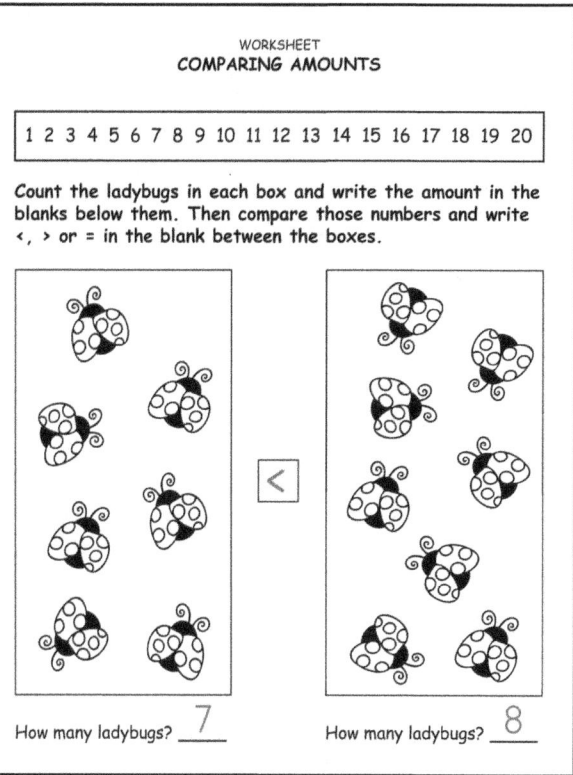

Page 38

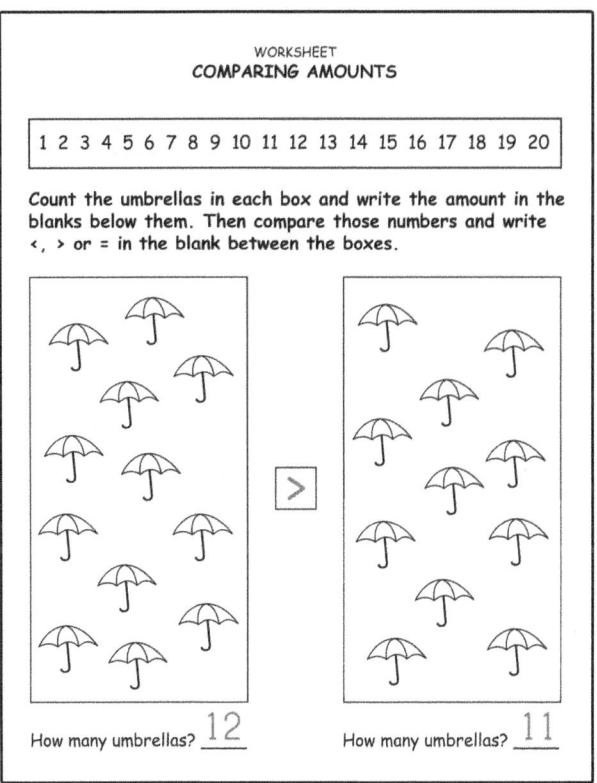

Page 39

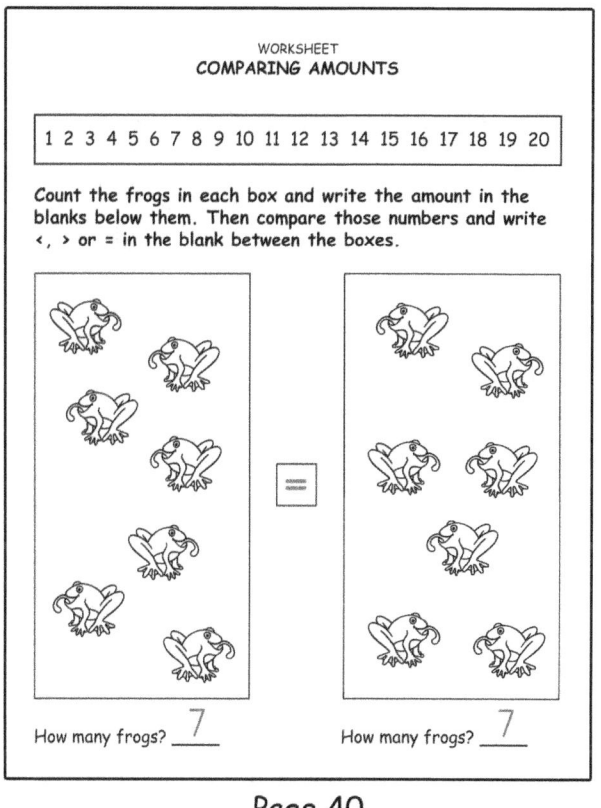

Page 40

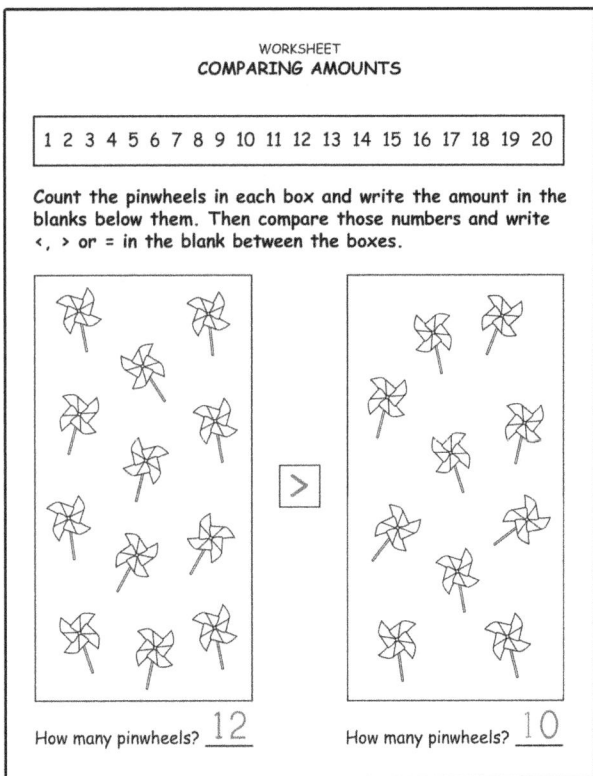
Page 41

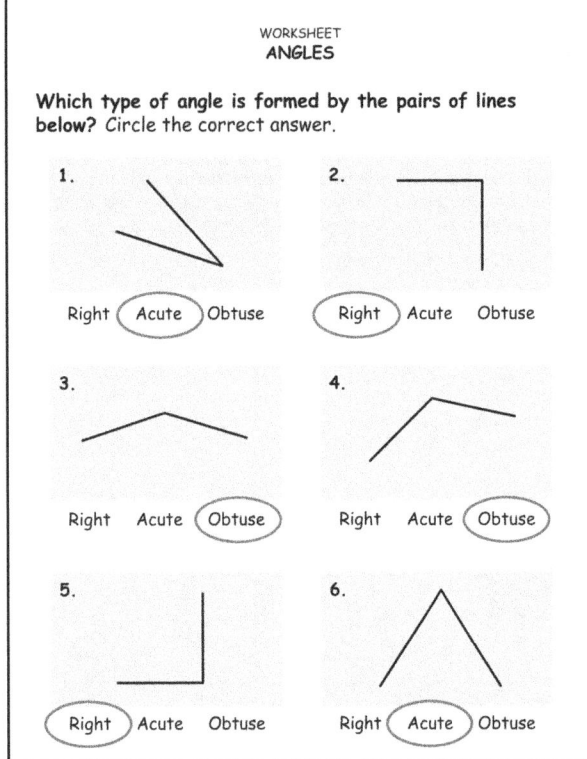
Page 49

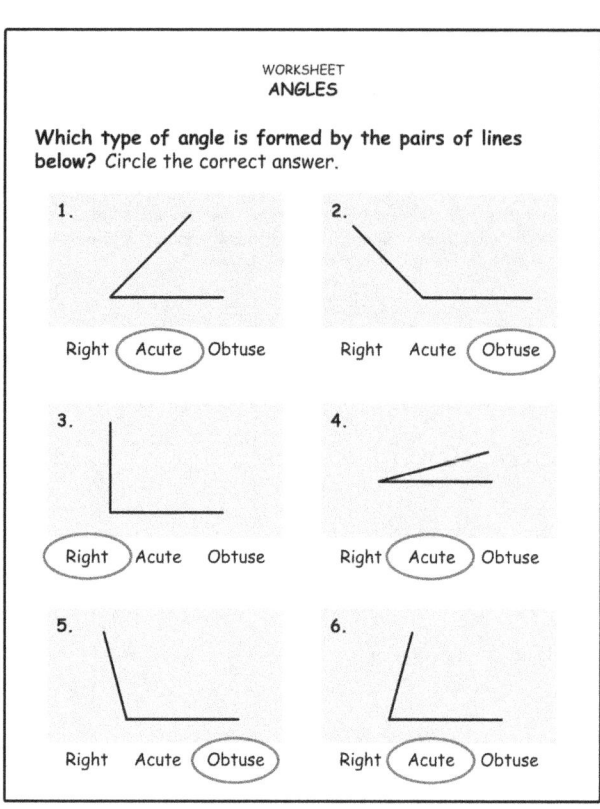
Page 48

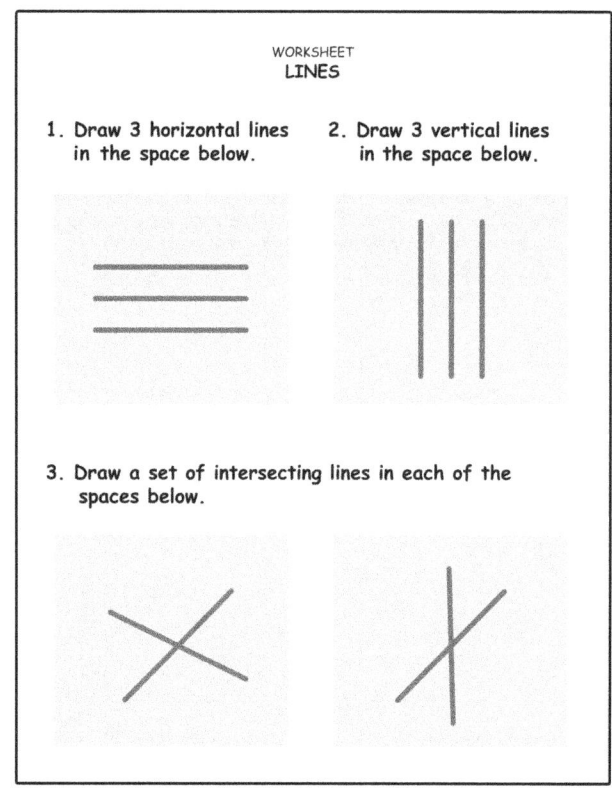
Page 59

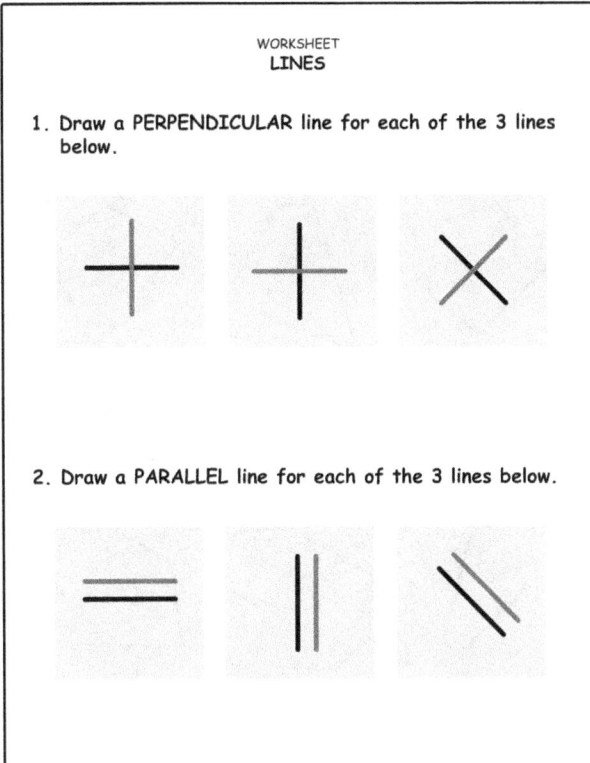

Page 60

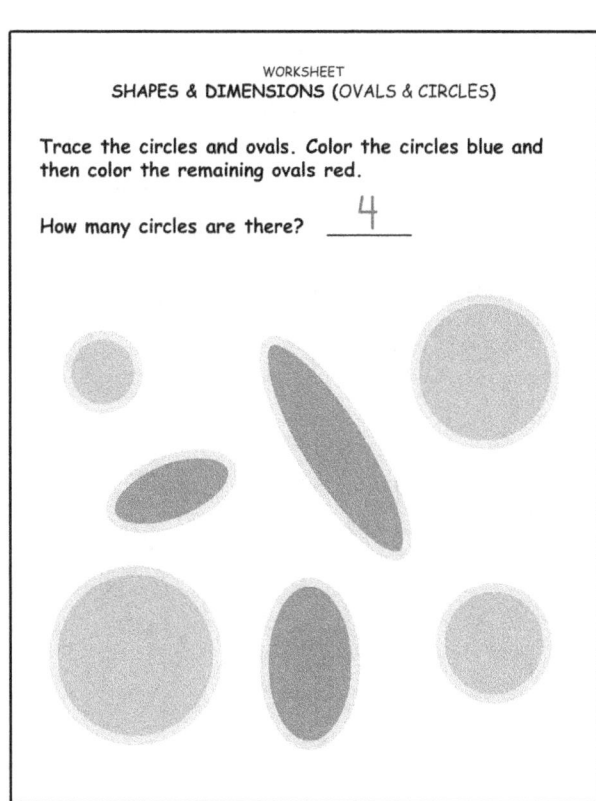

Page 75

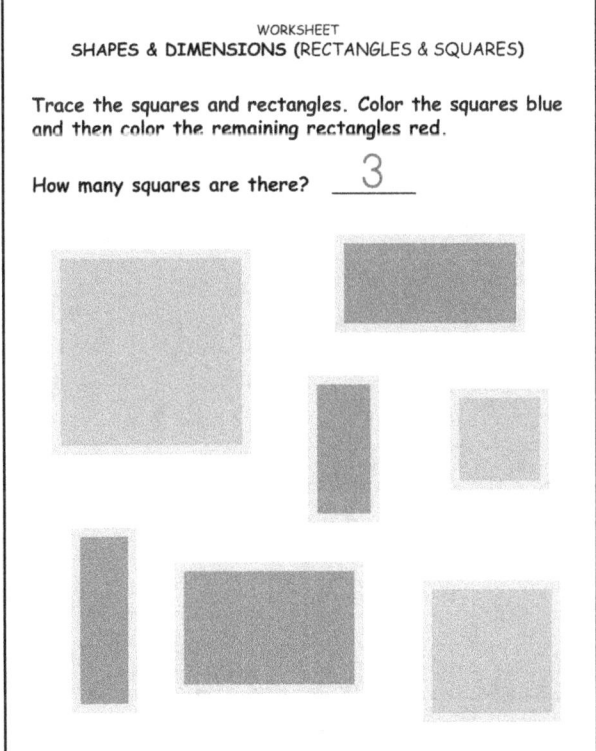

Page 76

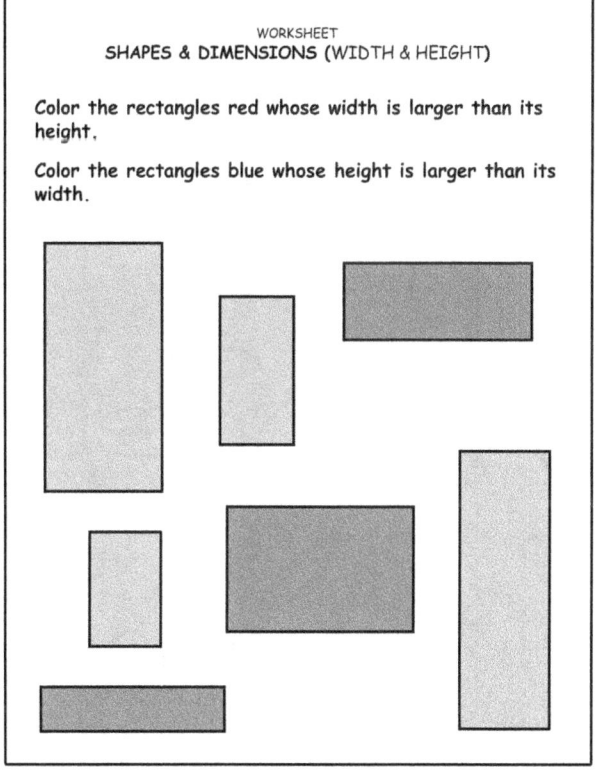

Page 77

WORKSHEET
SHAPES & DIMENSIONS (WIDTH & HEIGHT)

Color the ovals red whose width is larger than its height.
Color the ovals blue whose height is larger than its width.

Page 78

WORKSHEET
SHAPES & DIMENSIONS (VERTICES)

Count the number of vertices in each of the shapes and write the amounts in the blanks.

1. 5
2. 5
3. 6
4. 4
5. 8
6. 6

Page 79

WORKSHEET
SHAPES & DIMENSIONS (2-DIMENSIONAL SHAPES)

Count the different types of 2 dimensional shapes, then write the correct amounts in the blanks.

CIRCLES __4__   SQUARES __3__

OVALS __3__   RECTANGLES __2__

TRIANGLES __4__

Page 80

WORKSHEET
REGULAR POLYGONS

Count the number of sides and vertices and write the amounts in the blanks.

How many sides? 3

How many vertices? 3

Trace the shape name, then write it on your own in the space below:

equilateral triangle

equilateral triangle

Page 84

WORKSHEET
REGULAR POLYGONS

Count the number of sides and vertices and write the amounts in the blanks.

How many sides? 4
How many vertices? 4

Trace the shape name, then write it on your own in the space below:

square

square

Page 85

WORKSHEET
REGULAR POLYGONS

Count the number of sides and vertices and write the amounts in the blanks.

How many sides? 5
How many vertices? 5

Trace the shape name, then write it on your own in the space below:

pentagon

pentagon

Page 86

WORKSHEET
REGULAR POLYGONS

Count the number of sides and vertices and write the amounts in the blanks.

How many sides? 6
How many vertices? 6

Trace the shape name, then write it on your own in the space below:

hexagon

hexagon

Page 87

WORKSHEET
REGULAR POLYGONS

Count the number of sides and vertices and write the amounts in the blanks.

How many sides? 8
How many vertices? 8

Trace the shape name, then write it on your own in the space below:

octagon

octagon

Page 88

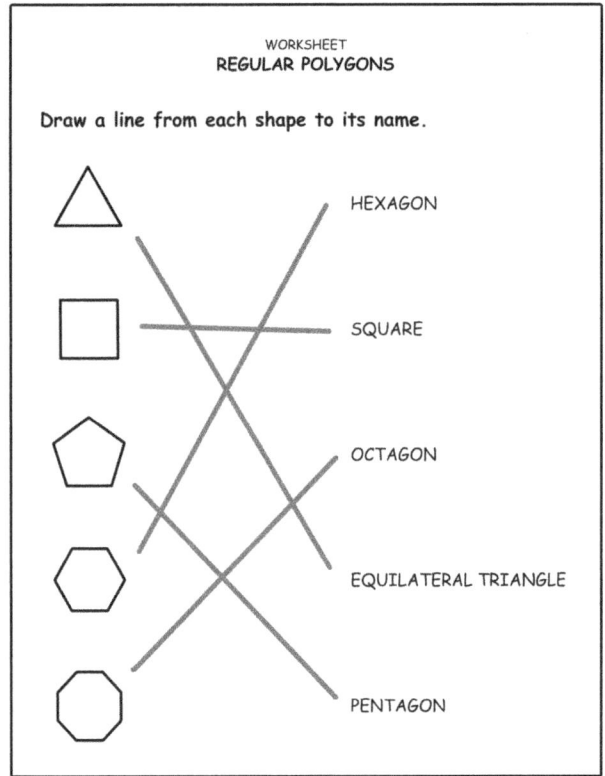

Page 89

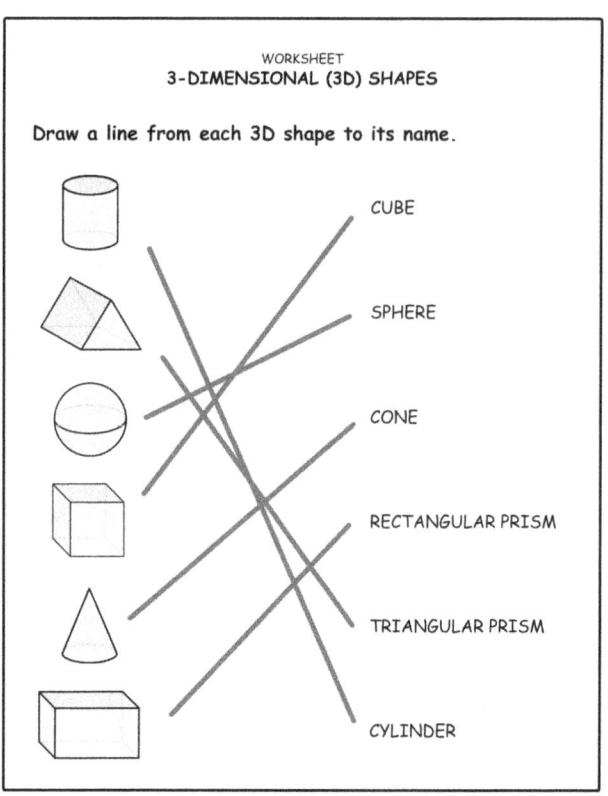

Page 100

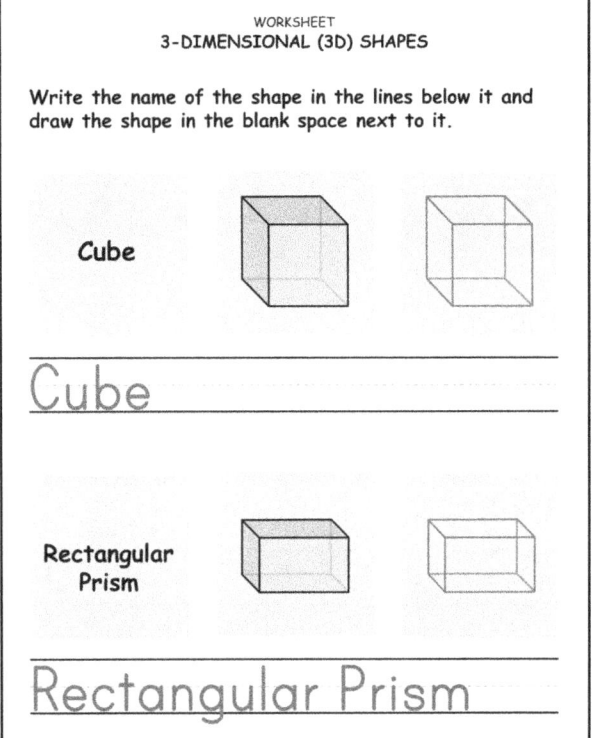

Page 101

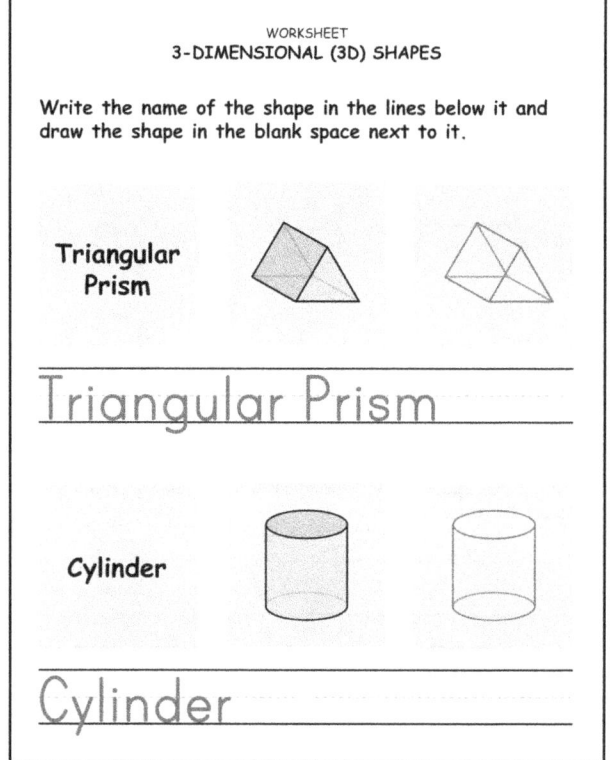

Page 102

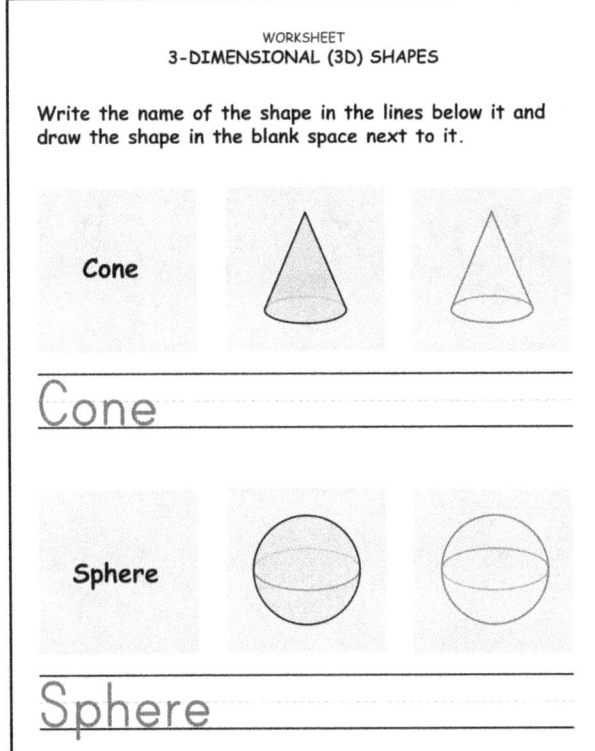

Page 103

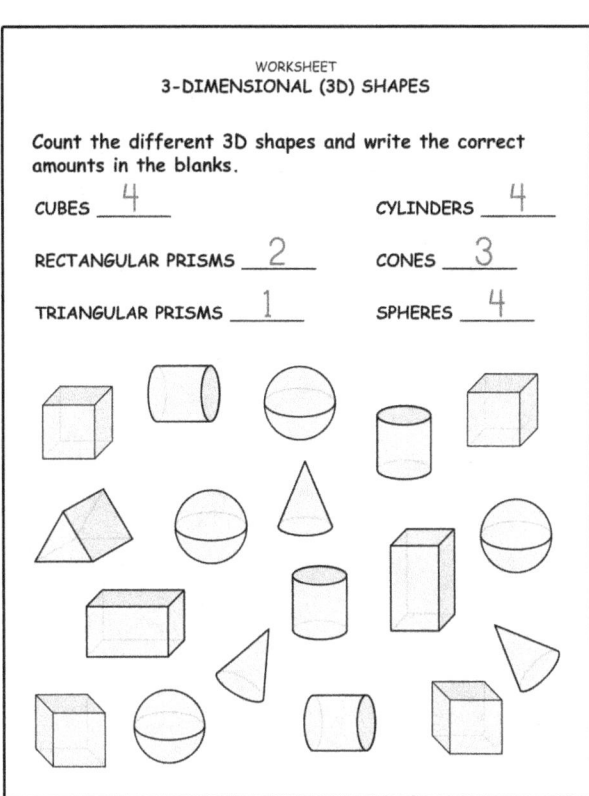

Page 104

WORKSHEET
ADDITION

1. Find the sum: 6 + 1
Trace and complete the addition statement below.
Color in 6 smilies, then color in 1 more smilie to help find the sum.

6 + 1 = 7

2. Find the sum: 5 + 2
Trace and complete the addition statement below.
Color in 5 beach balls, then color in 2 more beach balls to help find the sum.

5 + 2 = 7

Page 111

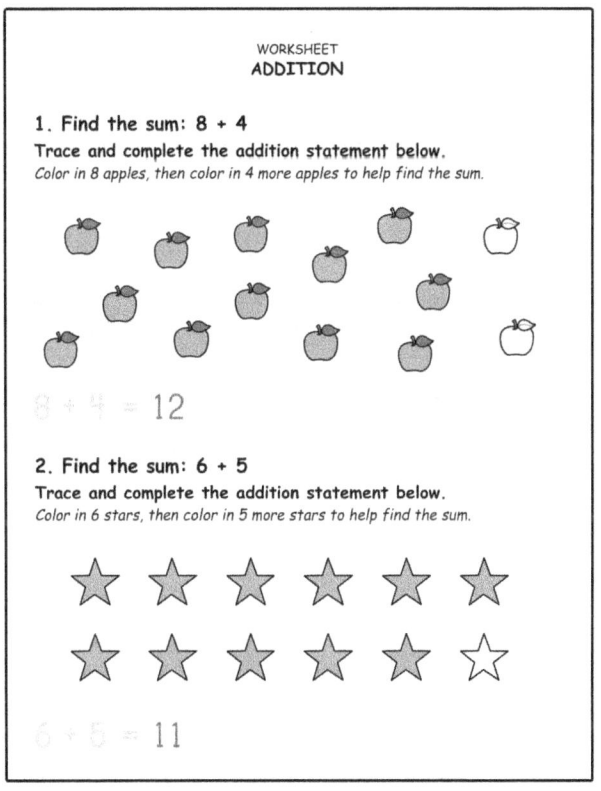

Page 112

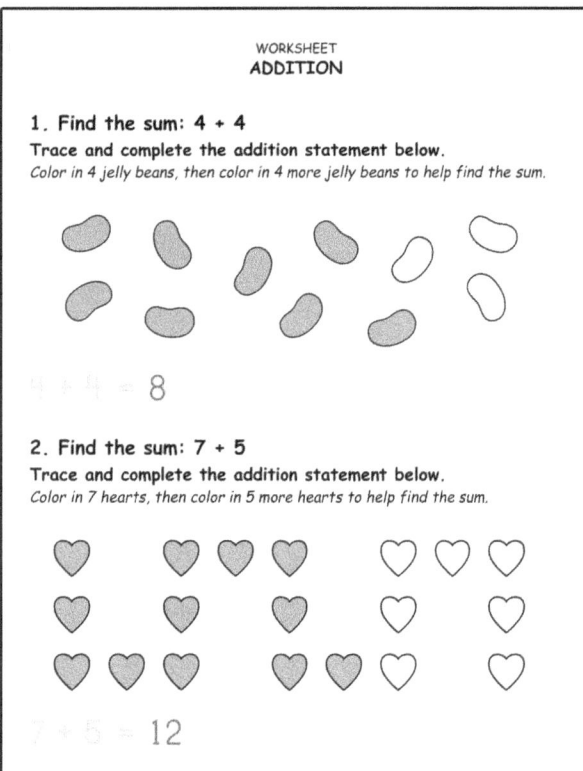

Page 113

WORKSHEET
ADDITION

1. Find the sum: 2 + 2 + 1
Trace and complete the addition statement below.
Color in 2 bowls, then 2 more bowls and another 1 bowl to help find the sum.

2 + 2 + 1 = 5

2. Find the sum: 3 + 2 + 3
Trace and complete the addition statement below.
Color in 3 triangles, then 2 more triangles and another 3 triangles to help find the sum.

3 + 2 + 3 = 8

Page 114

WORKSHEET
THE COMMUTATIVE PROPERTY

Using the commutative property of addition, rewrite each addition statement by switching the positions of the numbers being added.

2 + 3 = 5 → 3 + 2 = 5
1 + 3 = 4 → 3 + 1 = 4
3 + 5 = 8 → 5 + 3 = 8
1 + 4 = 5 → 4 + 1 = 5
6 + 7 = 13 → 7 + 6 = 13
7 + 8 = 15 → 8 + 7 = 15

Page 120

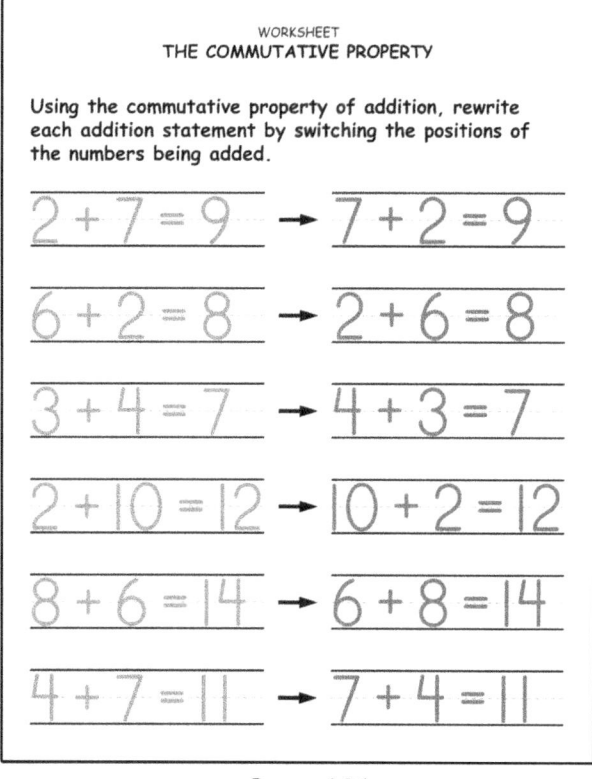

Page 121

WORKSHEET
THE COMMUTATIVE PROPERTY

Using the commutative property of addition, rewrite each addition statement by switching the positions of the numbers being added.

0 + 3 = 3	→ 3 + 0 = 3
1 + 7 = 8	→ 7 + 1 = 8
3 + 6 = 9	→ 6 + 3 = 9
6 + 4 = 10	→ 4 + 6 = 10
2 + 8 = 10	→ 8 + 2 = 10
4 + 8 = 12	→ 8 + 4 = 12

Page 122

WORKSHEET
SUBTRACTION

1. Find the difference: 4 - 1
Trace and complete the subtraction statement below.
Color in 4 hearts, then draw an X through 1 colored-in heart to remove it. Count the remaining colored-in hearts to find the difference.

4 - 1 = 3

2. Find the difference: 6 - 3
Trace and complete the subtraction statement below.
Color in 6 balls, then draw an X through 3 colored-in balls to remove them. Count the remaining colored-in balls to find the difference.

6 - 3 = 3

Page 129

WORKSHEET
SUBTRACTION

1. Find the difference: 5 - 4
Trace and complete the subtraction statement below.
Color in 5 stars, then draw an X through 4 colored-in stars to remove them. Count the remaining colored-in stars to find the difference.

5 - 4 = 1

2. Find the difference: 7 - 2
Trace and complete the subtraction statement below.
Color in 7 smilies, then draw an X through 2 colored-in smilies to remove them. Count the remaining colored-in smilies to find the difference.

7 - 2 = 5

Page 130

WORKSHEET
SUBTRACTION

1. Find the difference: 10 - 4
Trace and complete the subtraction statement below.
Color in 10 beach balls, then draw an X through 4 colored-in beach balls to remove them. Count the remaining colored-in balls to find the difference.

10 - 4 = 6

2. Find the difference: 12 - 5
Trace and complete the subtraction statement below.
Color in 12 jelly beans, then draw an X through 5 colored-in jelly beans to remove them. Count the remaining colored-in jelly beans to find the difference.

12 - 5 = 7

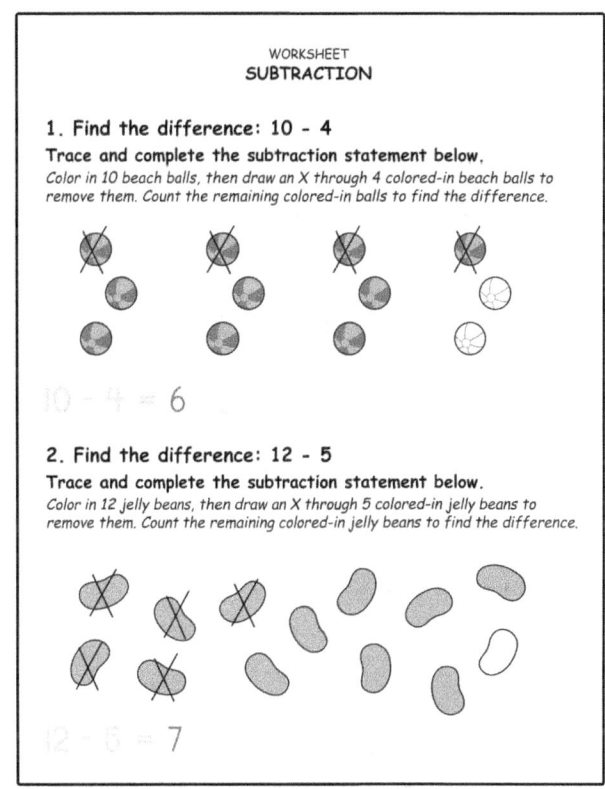

Page 131

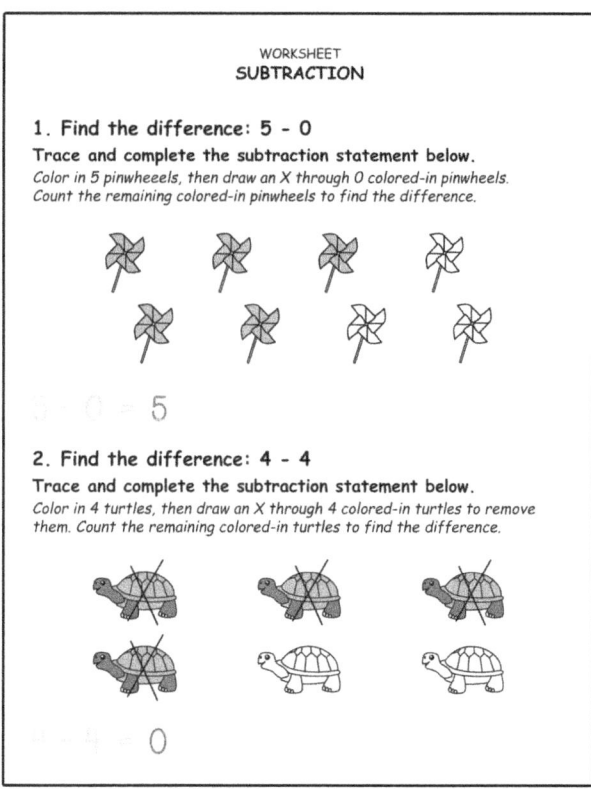

Page 132

WORKSHEET
TABLES: ROWS & COLUMNS

Use the table to answer the questions below.

	Lunch	Toys
Abby	🍕	🦆
Chris	🥣	🧸
Mark	🍐	🌀

Circle the correct answers.

1. Which lunch belongs to Abby?

2. Who has 🥣 for lunch?   Abby   (Chris)   Mark

3. Which toy belongs to Abby?

4. Whose toy is a 🌀 ?   Abby   Chris   (Mark)

Page 138

WORKSHEET
TABLES: ROWS & COLUMNS

Kevin, Tina and Mary each have a flower garden. The table below shows the amounts and types of flowers they are growing.

	🌷	🌻	🌼	🌱
Kevin	2	1	3	2
Tina	4	0	2	1
Mary	3	3	1	0

1. How many 🌻 does Mary have? _3_
2. How many 🌱 does Kevin have? _2_
3. How many 🌼 does Tina have? _2_
4. How many 🌻 does Tina have? _0_

Page 139

WORKSHEET
TABLES: ROWS & COLUMNS

Kevin, Tina and Mary each have a flower garden. The table below shows the amounts and types of flowers they are growing.

	🌷	🌻	🌼	🌱
Kevin	2	1	3	2
Tina	4	0	2	1
Mary	3	3	1	0

1. Who has 2 🌼 ? _Tina_
2. Who has 3 🌻 ? _Mary_
3. Who has no 🌱 ? _Mary_
4. Who has 2 🌷 ? _Kevin_

Page 140

WORKSHEET
TABLES: ROWS & COLUMNS

Kevin, Tina and Mary each have a flower garden. The table below shows the amounts and types of flowers they are growing.

	🌷	🌻	🌼	🌱
Kevin	2	1	3	2
Tina	4	0	2	1
Mary	3	3	1	0

1. How many 🌷 do Kevin and Tina have all together? __6__
2. How many flowers does Mary have in her garden? __7__
3. How many 🌱 do Kevin, Tina and Mary have all together? __3__

Page 141

WORKSHEET
TABLES: ROWS & COLUMNS

Kevin, Tina and Mary each have a flower garden. The table below shows the amounts and types of flowers they are growing.

	🌷	🌻	🌼	🌱
Kevin	2	1	3	2
Tina	4	0	2	1
Mary	3	3	1	0

1. How many 🌼 do Mary and Tina have all together? __3__
2. How many flowers does Kevin have in his garden? __8__
3. How many 🌷 do Kevin, Tina and Mary have all together? __9__

Page 142

WORKSHEET
COUNTING (UP TO 100)

Trace each group of numbers and fill in the blank with the number that comes AFTER them.

10 11 __12__   27 28 __29__
22 23 __24__   31 32 __33__
45 46 __47__   84 85 __86__
67 68 __69__   77 78 __79__
59 60 __61__   49 50 __51__
78 79 __80__   28 29 __30__

Page 152

WORKSHEET
COUNTING (UP TO 100)

Trace each group of numbers and fill in the blank with the number that comes BEFORE them.

__11__ 12 13   __16__ 17 18
__23__ 24 25   __40__ 41 42
__31__ 32 33   __81__ 82 83
__43__ 44 45   __66__ 67 68
__58__ 59 60   __28__ 29 30
__79__ 80 81   __69__ 70 71

Page 153

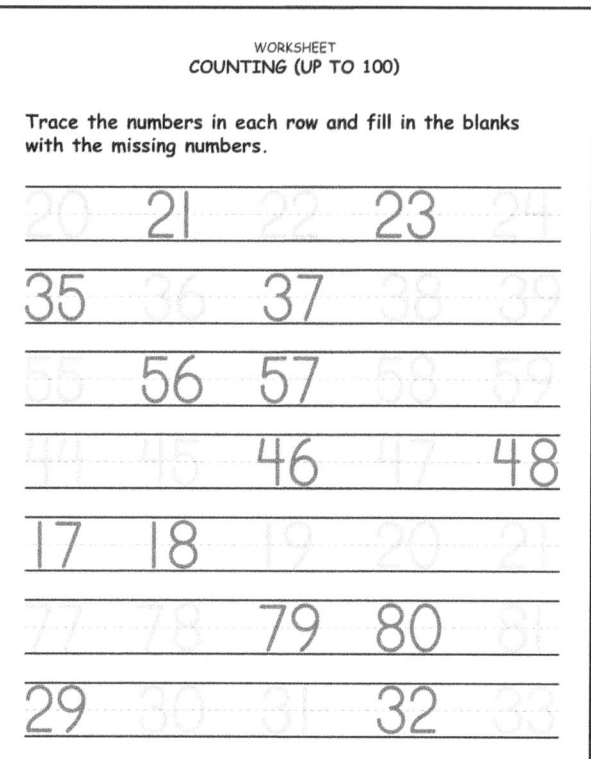

Page 154

Page 155

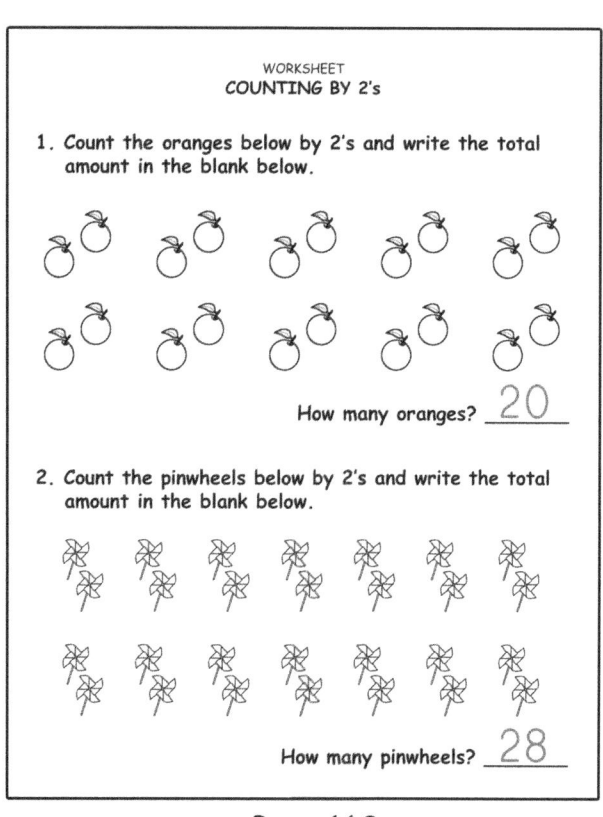

Page 161

Page 162

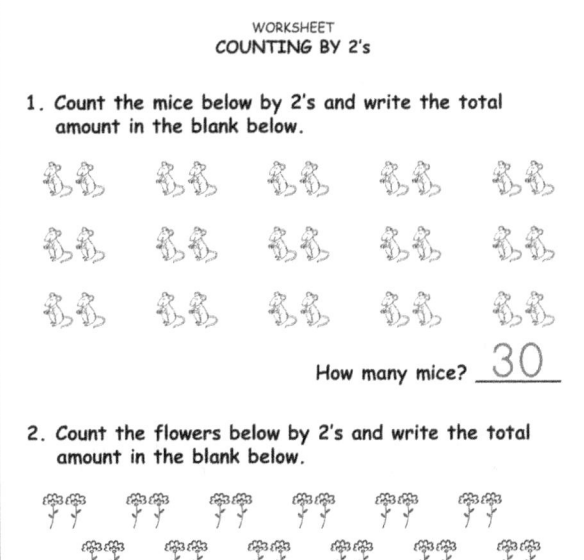

Page 163

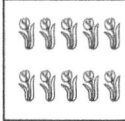 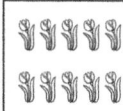

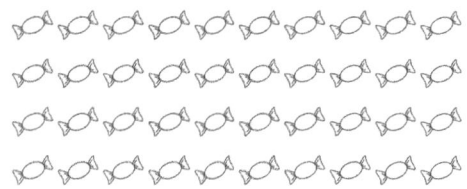

Page 169

Page 170

Page 171

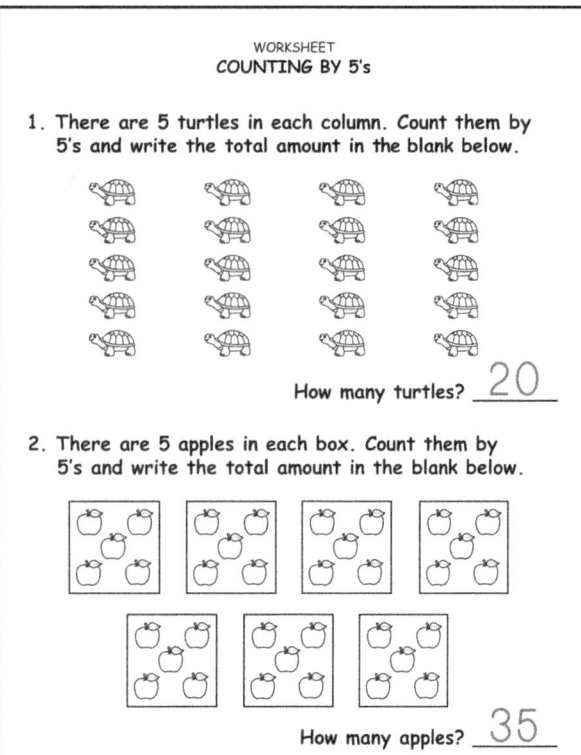

Page 177

WORKSHEET
COUNTING BY 5's

1. There are 5 fish in each fishbowl. Count them by 5's and write the total amount in the blank below.

   How many fish? __40__

2. There are 5 flowers in each group. Count them by 5's and write the total amount in the blank below.

   How many flowers? __50__

Page 178

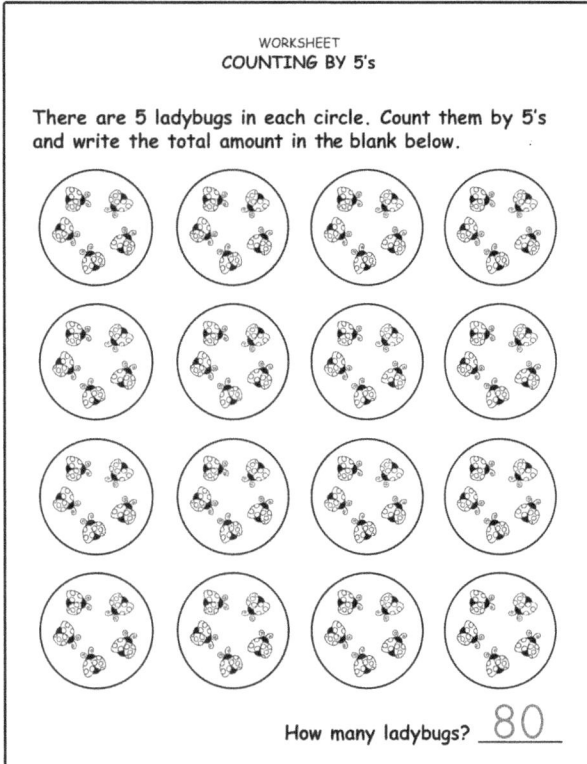

Page 179

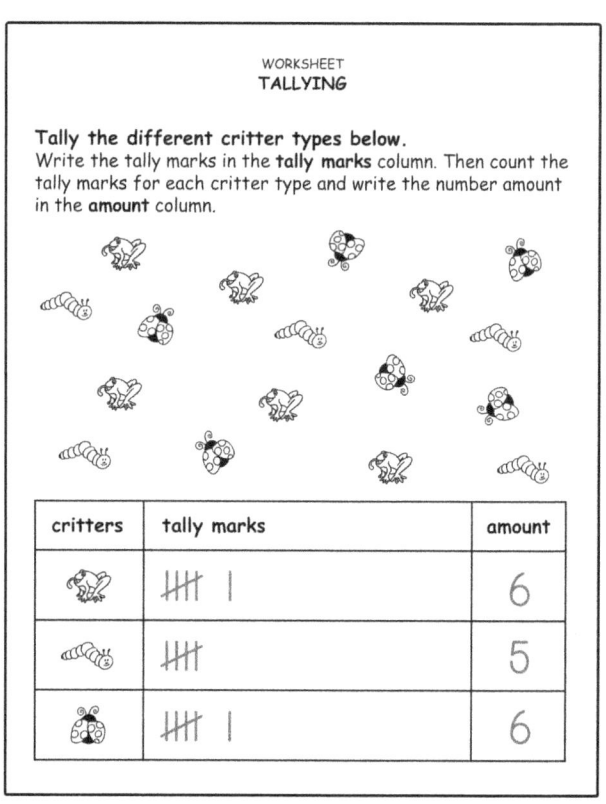

Page 185

Page 186

WORKSHEET
TALLYING

Tally the different animal types below.
Write the tally marks in the **tally marks** column. Then count the tally marks for each animal type and write the number amount in the **amount** column.

animals	tally marks	amount
(monkey)	IIII	4
(owl)	||||-I	6
(bird)	||||-II	7

Page 187

WORKSHEET
TALLYING

Tally the different flower types below.
Write the tally marks in the **tally marks** column. Then count the tally marks for each flower type and write the number amount in the **amount** column.

flowers	tally marks	amount
(tulip)	||||-| |||||-	10
(star)	||||-| |||	8
(flower)	||||-| |||	8

Page 188

WORKSHEET
TALLYING

Count the tally marks.
Below is a tally chart of favorite ice cream toppings. For each topping, count the tally marks and write the number amount in the amount column.

toppings	tally marks	amount
hot fudge	||||-| ||||-| ||||-| ||||-| ||||	24
caramel	||||-| ||||-| |||	13
whipped cream	||||-| ||||-| ||||-|		16

Which topping was favored most?
hot fudge

Which topping was favored least?
caramel

Page 189

WORKSHEET
TALLYING

Count the tally marks.
Below is a tally chart of favorite fruits. For each fruit, count the tally marks and write the number amount in the amount column.

fruit	tally marks	amount
orange	||||-| ||||-| ||||-| ||||-|	20
banana	||||-| ||||-| ||||	14
watermelon	||||-| ||||-| ||||-| ||||-| |||	23

Which fruit was favored most?
watermelon

Which fruit was favored least?
banana

WORKSHEET
TALLYING

Count the tally marks.
Below is a tally chart of favorite sports. For each sport, count the tally marks and write the number amount in the amount column.

sports	tally marks	amount
basketball	|||| |||| |||| ||||	20
baseball	|||| |||| |||| ||||	19
football	|||| |||| |||| |||| ||	22

Which sport was favored most?
football

Which sport was favored least?
baseball

www.ingramcontent.com/pod-product-compliance
Lightning Source LLC
Chambersburg PA
CBHW081347080526
44588CB00016B/2400